T0212238

The Datacenter as a Computer

Designing Warehouse-Scale Machines

Third Edition

Synthesis Lectures on Computer Architecture

Editor

Margaret Martonosi, *Princeton University*

Synthesis Lectures on Computer Architecture publishes 50- to 100-page books on topics pertaining to the science and art of designing, analyzing, selecting, and interconnecting hardware components to create computers that meet functional, performance, and cost goals. The scope will largely follow the purview of premier computer architecture conferences, such as ISCA, HPCA, MICRO, and ASPLOS.

The Datacenter as a Computer: Designing Warehouse-Scale Machines, Third Edition
Luiz André Barroso, Urs Hölzle, and Parthasarathy Ranganathan
2018

Principles of Secure Processor Architecture Design
Jakub Szefer
2018

General-Purpose Graphics Processor Architectures
Tor M. Aamodt, Wilson Wai Lun Fung, and Timothy G. Rogers
2018

Compiling Algorithms for Heterogeneous Systems
Steven Bell, Jing Pu, James Hegarty, and Mark Horowitz
2018

Architectural and Operating System Support for Virtual Memory
Abhishek Bhattacharjee and Daniel Lustig
2017

Deep Learning for Computer Architects
Brandon Reagen, Robert Adolf, Paul Whatmough, Gu-Yeon Wei, and David Brooks
2017

On-Chip Networks, Second Edition
Natalie Enright Jerger, Tushar Krishna, and Li-Shiuan Peh
2017

Space-Time Computing with Temporal Neural Networks
James E. Smith
2017

Hardware and Software Support for Virtualization No Access
Edouard Bugnion, Jason Nieh, and Dan Tsafrir
2017

Datacenter Design and Management: A Computer Architect's Perspective
Benjamin C. Lee
2016

A Primer on Compression in the Memory Hierarchy
Somayeh Sardashti, Angelos Arelakis, Per Stenström, and David A. Wood
2015

Research Infrastructures for Hardware Accelerators
Yakun Sophia Shao and David Brooks
2015

Analyzing Analytics
Rajesh Bordawekar, Bob Blainey, and Ruchir Puri
2015

Customizable Computing
Yu-Ting Chen, Jason Cong, Michael Gill, Glenn Reinman, and Bingjun Xiao
2015

Die-stacking Architecture
Yuan Xie and Jishen Zhao
2015

Single-Instruction Multiple-Data Execution
Christopher J. Hughes
2015

Power-Efficient Computer Architectures: Recent Advances
Magnus Själander , Margaret Martonosi, and Stefanos Kaxiras
2014

The Datacenter as a Computer: Designing Warehouse-Scale Machines, Third Edition
Luiz André Barroso, Urs Hölzle, and Parthasarathy Ranganathan

ISBN: 978-3-031-00633-3 paperback
ISBN: 978-3-031-01761-2 ebook
ISBN: 978-3-031-00058-4 hardcover
ISBN: 9781681734361 epub

DOI 10.1007/978-3-031-01761-2

A Publication in the Springer series
SYNTHESIS LECTURES ON ADVANCES IN AUTOMOTIVE TECHNOLOGY
Lecture #46
Series Editor: Margaret Martonosi, Princeton University

Series ISSN 1935-4185 Print 1935-4193 Electronic

The cover photograph is of Google's data center in Eemshaven, Netherlands.

The Datacenter as a Computer
Designing Warehouse–Scale Machines

Third Edition

Luiz André Barroso, Urs Hölzle, and Parthasarathy Ranganathan
Google LLC

SYNTHESIS LECTURES ON COMPUTER ARCHITECTURE #46

ABSTRACT

This book describes *warehouse-scale computers* (WSCs), the computing platforms that power cloud computing and all the great web services we use every day. It discusses how these new systems treat the *datacenter itself as one massive computer designed at warehouse scale*, with hardware and software working in concert to deliver good levels of internet service performance. The book details the architecture of WSCs and covers the main factors influencing their design, operation, and cost structure, and the characteristics of their software base. Each chapter contains multiple real-world examples, including detailed case studies and previously unpublished details of the infrastructure used to power Google's online services. Targeted at the architects and programmers of today's WSCs, this book provides a great foundation for those looking to innovate in this fascinating and important area, but the material will also be broadly interesting to those who just want to understand the infrastructure powering the internet.

The third edition reflects four years of advancements since the previous edition and nearly doubles the number of pictures and figures. New topics range from additional workloads like video streaming, machine learning, and public cloud to specialized silicon accelerators, storage and network building blocks, and a revised discussion of data center power and cooling, and uptime. Further discussions of emerging trends and opportunities ensure that this revised edition will remain an essential resource for educators and professionals working on the next generation of WSCs.

KEYWORDS

data centers, cloud computing, servers, hyperscale systems, hardware accelerators, Internet services, distributed systems, energy efficiency, fault-tolerant computing, cluster computing, computer organization, computer design

Contents

Acknowledgments

While we draw from our direct involvement in Google's infrastructure design and operation over the past several years, most of what we have learned and now report here is the result of the hard work, insights, and creativity of our colleagues at Google. The work of our Technical Infrastructure teams directly supports the topics we cover here and, therefore, we are particularly grateful to them for allowing us to benefit from their experience.

Thanks to the work of Kristin Berdan at Google and Michael Morgan at Morgan & Claypool, we were able to make this lecture available electronically without charge, which was a condition for our accepting this task. We are very grateful to Mark Hill and Michael Morgan for inviting us to this project, and to Michael and Margaret Martonosi for the encouragement and gentle prodding and endless patience to make this revision possible.

Special thanks to several people at Google who helped review and added content in the Third Edition: Jichuan Chang, Liqun Cheng, Mike Dahlin, Tim Hockin, Dave Landhuis, David Lo, Beckett Madden, Jon McCune, Tipp Moseley, Nishant Patil, Alex Ramirez, John Stanley, Varun Sakalkar, and Amin Vahdat. Thanks to Robert Stroud, Andrew Morgan, and others who helped with data and figures for prior editions. Special thanks to Jimmy Clidaras who wrote the original content on cooling and power distribution for Chapter 5 of the second edition. Ana Lopez-Reynolds, along with Marcos Armenta, Gary Borella, and Loreene Garcia helped with all the illustrations, creating awesome new figures, and making sure all the illustrations had a consistent look. Thanks to our reviewers: Christos Kozyrakis and Thomas Wenisch for their valuable comments, and to Bob Silva for his proofreading. Thanks also to Brent Beckley, Deborah Gabriel, and Christine Kiilerich at Morgan Claypool for their collaboration.

The second edition greatly benefited from thorough reviews by David Andersen, Partha Ranganathan, and Christos Kozyrakis, as well as contributions and corrections by Tor Aamodt, Dilip Agrawal, Remzi Arpaci-Dusseau, Mike Bennett, Liqun Chen, Xiaobo Fan, David Guild, Matthew Harris, Mark Hennecke, Mark Hill, Thomas Olavson, Jack Palevich, Pete Pellerzi, John Reese, Ankit Somani, and Amin Vahdat. We sincerely thank them for their help. We are also grateful for the feedback and corrections to the first edition submitted by Vijay Rao, Robert Hundt, Mike Marty, David Konerding, Jeremy Dion, Juan Vargas, Artur Klauser, Pedro Reviriego Vasallo, Amund Tveit, Xiau Yu, Bartosz Prybylski, Laurie Doyle, Marcus Fontoura, Steve Jenkin, Evan Jones, Chuck Newman, Taro Tokuhiro, Jordi Torres, and Christian Belady. Ricardo Bianchini, Fred Chong, Jeff Dean, and Mark Hill provided extremely useful feedback on early drafts despite being handed a relatively immature early version of the first edition. The first edition also benefited from proofreading by Catherine Warner.

And finally, we appreciate you, our readers, for all the support for this book and all the feedback on the two prior editions. We will continue to much appreciate any thoughts, suggestions, or corrections you might have on this edition as well. Please submit your comments and errata at https://goo.gl/HHqQ25. Thanks in advance for taking the time to contribute.

CHAPTER 1

Introduction

The ARPANET is nearly 50 years old, and the World Wide Web is close to 3 decades old. Yet the internet technologies that emerged from these two remarkable milestones continue to transform industries and cultures today and show no signs of slowing down. The widespread use of such popular internet services as web-based email, search, social networks, online maps, and video streaming, plus the increased worldwide availability of high-speed connectivity, have accelerated a trend toward server-side or "cloud" computing. With such trends now embraced by mainline enterprise workloads, the cloud computing market is projected to reach close to half a trillion dollars in the next few years [Col17].

In the last few decades, computing and storage have moved from PC-like clients to smaller, often mobile, devices, combined with large internet services. While early internet services were mostly informational, today, many web applications offer services that previously resided in the client, including email, photo and video storage, and office applications. Increasingly, traditional enterprise workloads are also shifting to cloud computing. This shift is driven not only by the need for user experience improvements, such as ease of management (no configuration or backups needed) and ubiquity of access, but also by the advantages it provides to vendors. Specifically, software as a service allows faster application development because it is simpler for software vendors to make changes and improvements. Instead of updating many millions of clients (with a myriad of peculiar hardware and software configurations), vendors need to only coordinate improvements and fixes inside their data centers and can restrict their hardware deployment to a few well-tested configurations. Similarly, server-side computing allows for faster introduction of new hardware innovations, such as accelerators that can be encapsulated under well-defined interfaces and APIs. Moreover, data center economics allow many application services to run at a low cost per user. For example, servers may be shared among thousands of active users (and many more inactive ones), resulting in better utilization. Similarly, the computation and storage itself may become cheaper in a shared service (for example, an email attachment received by multiple users can be stored once rather than many times, or a video can be converted to a client format once and streamed to multiple devices). Finally, servers and storage in a data center can be easier to manage than the desktop or laptop equivalent because they are under control of a single, knowledgeable entity. Security, in particular, is an important differentiator for the cloud.

Some workloads require so much computing capability that they are a more natural fit for a massive computing infrastructure than for client-side computing. Search services (web, images, and so on) are a prime example of this class of workloads, but applications such as language translation (and more broadly, machine learning) can also run more effectively on large shared computing installations because of their reliance on massive-scale models.

These trends toward server-side computing and widespread internet services created a new class of computing systems that, in our first edition of this book, we named *warehouse-scale computers*, or WSCs. The name is meant to call attention to the most distinguishing feature of these machines: the *massive scale* of their software infrastructure, data repositories, and hardware platform. This perspective is a departure from a view of the computing problem that implicitly assumes a model where one program runs in a single machine. In warehouse-scale computing, the program is an internet service, which may consist of tens or more individual programs that interact to implement complex end-user services such as email, search, or maps. These programs might be implemented and maintained by different teams of engineers, perhaps even across organizational, geographic, and company boundaries (as is the case with mashups, for example).

The computing platform required to run such large-scale services bears little resemblance to a pizza-box server or even the refrigerator-sized, high-end multiprocessors that reigned in prior decades. The hardware for such a platform consists of thousands of individual computing nodes with their corresponding networking and storage subsystems, power distribution and conditioning equipment, and extensive cooling systems. The enclosure for these systems is in fact a building, which is often indistinguishable from a large warehouse.

1.1   WAREHOUSE-SCALE COMPUTERS

Had scale been the only distinguishing feature of these systems, we might simply refer to them as *data centers*. Data centers are buildings where multiple servers and communication gear are co-located because of their common environmental requirements and physical security needs, and for ease of maintenance. In that sense, a WSC is a type of data center. Traditional data centers, however, typically host a large number of relatively small- or medium-sized applications, each running on a dedicated hardware infrastructure that is de-coupled and protected from other systems in the same facility. Those data centers host hardware and software for multiple organizational units or even different companies. Different computing systems within such a data center often have little in common in terms of hardware, software, or maintenance infrastructure, and tend not to communicate with each other at all.

WSCs currently power the services offered by companies such as Google, Amazon, Facebook, Microsoft, Alibaba, Tencent, Baidu, and others. They differ significantly from traditional data centers: they belong to a single organization, use a relatively homogeneous hardware and system software platform, and share a common systems management layer. Often, much of the application, middleware, and system software is built in-house compared to the predominance of third-party software running in conventional data centers. Most importantly, WSCs run a smaller number of very large applications (or internet services), and the common resource management infrastructure allows significant deployment flexibility. The requirements of homogeneity, single-organization control, and enhanced focus on cost efficiency motivate designers to take new approaches in constructing and operating these systems.

Although initially designed for online data-intensive web workloads, WSCs also now power public clouds such as those from Amazon, Google, and Microsoft. Such public clouds do run many small applications, like a regular data center. However, as seen by the provider, all of these applications are identical VMs, and they access large, common services for block or database storage, load balancing, and so on, fitting very well with the WSC model. Over the years, WSCs have also adapted to incorporate industry standards and more general purpose designs that the same infrastructure can support both large online internal services as well as large public cloud offerings, with very little deviation, and in many cases, even the same developer experience.

Internet services must achieve high availability, typically aiming for at least 99.99% uptime ("four nines," or about an hour of downtime per year). Achieving fault-free operation on a large collection of hardware and system software is hard and is made more difficult by the large number of servers involved. Although it might be theoretically possible to prevent hardware failures in a collection of 10,000 servers, it would surely be very expensive. Consequently, WSC workloads must be designed to gracefully tolerate large numbers of component faults with little or no impact on service level performance and availability.

1.2   COST EFFICIENCY AT SCALE

Building and operating a large computing platform is expensive, and the quality of a service it provides may depend on the aggregate processing and storage capacity available, further driving costs up and requiring a focus on cost efficiency. For example, in information retrieval systems such as web search, three main factors drive the growth of computing needs.

- Increased service popularity translates into higher request loads.

- The size of the problem keeps growing—the web is growing by millions of pages per day, which increases the cost of building and serving a web index.

- Even if the throughput and data repository could be held constant, the competitive nature of this market continuously drives innovations to improve the quality of results retrieved and the frequency with which the index is updated. Although some quality improvements can be achieved by smarter algorithms alone, most substantial improvements demand additional computing resources for every request. For example, in a search system that also considers synonyms of the search terms in a query, or semantic relationships, retrieving results is substantially more expensive. Either the search needs to retrieve documents that match a more complex query that includes the synonyms, or the synonyms of a term need to be replicated in the index data structure for each term.

This relentless demand for more computing capabilities makes cost efficiency a primary metric of interest in the design of WSCs. Cost efficiency must be defined broadly to account for all the significant components of cost, including hosting-facility capital and operational expenses (which include power provisioning and energy costs), hardware, software, management personnel, and repairs. Chapter 6 discusses these issues in more detail.

1.3   NOT JUST A COLLECTION OF SERVERS

Our central point is that the data centers powering many of today's successful internet services are no longer simply a collection of miscellaneous machines co-located in a facility and wired up together. The software running on these systems, such as Gmail or web search services, executes at a scale far beyond a single machine or a single rack: it runs on no smaller a unit than clusters of hundreds to thousands of individual servers. Therefore, the machine, the computer, is itself this large cluster or aggregation of servers and needs to be considered as a single computing unit.

The technical challenges of designing WSCs are no less worthy of the expertise of computer systems architects than any other class of machines. First, they are a new class of large-scale machines driven by a new and rapidly evolving set of workloads. Their size alone makes them difficult to experiment with or simulate efficiently; therefore, system designers must develop new techniques to guide design decisions. In addition, fault behavior, security, and power and energy considerations have a more significant impact in the design of WSCs, perhaps more so than in other smaller scale computing platforms. Finally, WSCs have an additional layer of complexity beyond systems consisting of individual servers or small groups of servers; WSCs introduce a significant new challenge to programmer productivity, a challenge even greater than programming the individual multicore systems that comprise the WSC. This additional complexity arises indirectly from virtualization and the larger scale of the application domain and manifests itself as a deeper and less homogeneous storage hierarchy (Chapter 4), higher fault rates (Chapter 7), higher performance variability (Chapter 2), and greater emphasis on microsecond latency tolerance (Chapter 8).

The objectives of this book are to introduce this new design space, describe some of the requirements and characteristics of WSCs, highlight some of the important challenges unique to this space, and share some of our experience designing, programming, and operating them within Google. We are fortunate to be not only designers of WSCs but also customers and programmers of the platform, which has provided us an unusual opportunity to evaluate design decisions throughout the lifetime of a product. We hope that we succeed in relaying our enthusiasm for this area as an exciting new target worthy of the attention of the general research and technical communities.

1.4   ONE DATA CENTER VS. SEVERAL

In this book, we define the computer to be architected as one data center, even though many internet services use multiple data centers located far apart. Multiple data centers are sometimes used as complete replicas of the same service, with replication used mostly to reduce user latency and improve serving throughput. In those cases, a given user request is fully processed within one data center, and our machine definition seems appropriate.

In cases where a user query involves computation across multiple data centers, our single data center focus is a less obvious fit. Typical examples are services that deal with nonvolatile user data updates, and therefore require multiple copies for disaster tolerance. For such computations, a set of data centers might be the more appropriate system. Similarly, video streaming workloads benefit significantly from a content-distribution network (CDN) across multiple data centers and edge points.

However, we have chosen to think of the multi-data center scenario as more analogous to a network of computers. This is in part to limit the scope of this book, but also mainly because the huge gap in connectivity quality between intra- and inter-data center communications causes programmers to view such systems as separate computational resources. As the software development environment for this class of applications evolves, or if the connectivity gap narrows significantly in the future, we may need to adjust our choice of machine boundaries.

1.5   WHY WSCS MIGHT MATTER TO YOU

In the first edition of the book a decade ago, we discussed how WSCs might be considered a niche area, because their sheer size and cost render them unaffordable by all but a few large internet companies. We argued then that we did not believe this to be true, and that the problems that large internet services face would soon be meaningful to a much larger constituency because many organizations would be able to afford similarly sized computers at a much lower cost.

Since then our intuition has come true. Today, the attractive economics of low-end server class computing platforms puts clusters of thousands of nodes within the reach of a relatively broad range of corporations and research institutions. Combined with the trends around ever increasing numbers of processor cores on a single die, a single rack of servers today has more hardware threads than many data centers had a decade ago. For example, a rack with 40 servers, each with four 16-core dual-threaded CPUs, contains more than 4,000 hardware threads! Such systems exhibit the same scale, architectural organization, and fault behavior of WSCs from the last decade.[1]

---

[1] The relative statistics about sources of hardware faults can change substantially in these more integrated future systems, but silicon trends around less reliable components and the likely continuing high impact of software-driven faults suggest that programmers of such systems still need to deal with a fault-ridden platform.

However, perhaps more importantly, the explosive growth and popularity of Infrastructure-as-a-Service (IaaS) cloud computing offerings have now made WSCs available to anyone with a credit card [Arm+10]. We believe that our experience building these systems is useful in understanding the design issues and programming challenges for the next-generation cloud computing platform.

1.6   ARCHITECTURAL OVERVIEW OF WSCS

The hardware implementation of a WSC differs significantly from one installation to the next. Even within a single organization such as Google, systems deployed in different years use different basic elements, reflecting the hardware improvements provided by the industry. However, the architectural organization of these systems is relatively stable. Therefore, it is useful to describe this general architecture at a high level as it sets the background for subsequent discussions.

1.6.1   SERVERS

The hardware building blocks for WSCs are low-end servers, typically in a 1U^2 or blade enclosure format, and mounted within a rack and interconnected using a local Ethernet switch. These rack-level switches, which can use 40 Gbps or 100 Gbps links, have a number of uplink connections to one or more cluster-level (or data center-level) Ethernet switches. This second-level switching domain can potentially span more than 10,000 individual servers. In the case of a blade enclosure, there is an additional first level of networking aggregation within the enclosure where multiple processing blades connect to a small number of networking blades through an I/O bus such as PCIe. Figure 1.1(a) illustrates a Google server building block. More recently, WSCs have featured additional compute hardware building blocks, including GPUs and custom accelerators (for example, Figure 1.1(b) illustrates a TPU board). Similar to servers, these are connected through custom or industry-standard interconnects at the rack (or multi-rack *pod*) levels, leading up to the data center network. Figure 1.1(c) illustrates storage trays that are used to build out storage. Figure 1.2 shows how these building blocks are assembled into rows of racks of servers for both general-purpose servers and accelerators.

---

^{2} Being satisfied with neither the metric nor the U.S. system, rack designers use "rack units" to measure the height of servers. 1U is 1.75 in or 44.45 mm; a typical rack is 42U high.

Figure 1.1: Example hardware building blocks for WSCs. Left to right: (a) a server board, (b) an accelerator board (Google's Tensor Processing Unit [TPU]), and (c) a disk tray.

Figure 1.2: Hardware building blocks assembled into interconnected racks and rows.

1.6.2   STORAGE

Disks and Flash SSDs are the building blocks of today's WSC storage system. To provide durable storage to a large number of applications, these devices are connected to the data center network and managed by sophisticated distributed systems. WSC system designers need to make several tradeoffs based on their requirements. For example, should the disks and SSDs be connected directly to compute servers (Directly Attached Storage, or DAS) or disaggregated as part of Network Attached Storage (NAS)? While DAS can reduce hardware costs and improve network utilization (the network port is dynamically shared between compute and storage tasks), the NAS approach tends to simplify deployment and provide higher QoS to avoid performance interference from compute jobs. Alos, WSCs, including Google's, deploy desktop-class disk drives (or their close cousins, nearline drives) instead of enterprise-grade disks to save costs. Overall, storage devices in WSCs should aim at an aggregated view of the global optima across key metrics, including bandwidth, IOPS, capacity, tail latency, and TCO.

Distributed storage systems not only manage storage devices, but also provide unstructured and structured APIs for application developers. Google's Google File System (GFS), and later Colossus and its Cloud cousin GCS [Ser17], are examples of unstructured WSC storage systems that use space-efficient Reed-Solomon codes and fast reconstruction for high availability. Google's BigTable [Cha+06] and Amazon's Dynamo [DeC+07] are examples of structured WSC storage that provides database-like functionality but with weaker consistency models. To simplify developers' tasks, newer generations of structured storage systems such as Spanner [Cor+12] provide an SQL-like interface and strong consistency models.

The nature of distributed storage in WSCs also leads to the interplay of storage and networking technologies. The fast evolution and improvement of data center networking have created a large gap between network and disk performance, to the point that WSC designs can be dramatically simplified to not consider disk locality. On the other hand, low-latency devices such as Flash SSD and emerging Non-Volatile Memories (NVMs) pose new challenges for WSC design.

WSC designers need to build balanced systems with a hierarchy of memory and storage technologies, holistically considering the cluster-level aggregate capacity, bandwidth, and latency. Chapter 3 discusses system balance in more detail.

1.6.3   NETWORKING FABRIC

Choosing a networking fabric for WSCs involves a trade-off between speed, scale, and cost. As of 2018, it is not hard to find switches with 48 ports to interconnect servers at full speed of 40–100 Gbps Ethernet within a single rack. As a result, bandwidth within a rack of servers tends to be homogeneous. However, network switches with high port counts, which are needed to tie together WSC clusters, have a much different price structure and are more than 10 times more expensive (per port) than commodity rack switches. As a rule of thumb, a switch with 10 times the bisection bandwidth often costs about 100 times more. As a result of this cost discontinuity, the networking fabric of WSCs is often organized as a two-level hierarchy. Commodity switches in each rack provide a fraction of their bisection bandwidth for inter-rack communication through a handful of uplinks to the more costly cluster-level switches. For example, a 48-port rack-level switch could connect 40 servers to 8 uplinks, for a 5:1 oversubscription (8–20 Gbps per server uplink bandwidth). In this network, programmers must be aware of the relatively scarce cluster-level bandwidth resources and try to exploit rack-level networking locality, complicating software development and possibly impacting resource utilization.

Alternatively, one can remove some of the cluster-level networking bottlenecks by spending more money on the interconnect fabric. For example, Infiniband interconnects typically scale to a few thousand ports but can cost much more than commodity Ethernet on a per port basis. Similarly, some networking vendors are starting to provide larger-scale Ethernet fabrics, but again at a much higher cost per server. How much to spend on networking vs. additional servers or storage is an application-specific question that has no single correct answer. However, for now, we will assume that intra-rack connectivity is cheaper than inter-rack connectivity.

1.6.4   BUILDINGS AND INFRASTRUCTURE

So far, we have discussed the compute, storage, and network building blocks of a WSC. These are akin to the CPU, memory, disk, and NIC components in a PC. We still need additional components like power supplies, fans, motherboards, chassis, and other components, to make a full computer. Similarly, a WSC has other important components related to power delivery, cooling, and building infrastructure that also need to be considered.

WSC buildings (and campuses) house the computing, network, and storage infrastructure discussed earlier, and design decisions on the building design can dramatically influence the availability and uptime of the WSC. (Chapter 4 discusses the tier levels used in the data center construction industry.)

Similarly, WSCs have elaborate power delivery designs. At the scale that they operate, WSCs can often consume more power than thousands of individual households. Therefore, they use a holistic and hierarchical power delivery design that feeds electricity from the utility, to the substation, to power distribution units, to bus ducts, to individual power rails and voltage regulators on the server board, while also providing corresponding backup and redundancy such as uninterruptible power supplies (UPSs), generators, and backup batteries at different levels of the topology.

WSCs can also generate a lot of heat. Similar to power delivery, WSCs employ an elaborate end-to-end cooling solution with a hierarchy of heat-exchange loops, from circulated air cooled by fan coils on the data center floor, to heat exchangers and chiller units, all the way to cooling towers that interact with the external environment.

The building design, delivery of input energy, and subsequent removal of waste heat all drive a significant fraction of data center costs proportional to the amount of power delivered, and also have implications on the design and performance of the compute equipment (for example, liquid cooling for accelerators) as well as the availability service level objectives (SLOs) seen by workloads. These are therefore as important to optimize as the design of the individual compute, storage, and networking blocks.

Figure 1.3: Power distribution, Council Bluffs, Iowa, U.S.

Figure 1.4: Data center cooling, Douglas County, Georgia, U.S.

Figure 1.5: Cooling towers and water storage tanks, Lenoir, North Carolina, U.S.

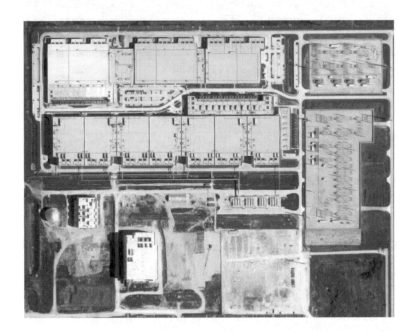

Figure 1.6: Aerial view of Google data center, Council Bluffs, Iowa, U.S.

Figure 1.7: Google Cloud Platform regions and number of zones, circa July 2018. The latest source is available at https://cloud.google.com/about/locations/.

1.6.5   POWER USAGE

Energy and power usage are important concerns in the design of WSCs because, as we discuss in more detail in Chapter 5, energy-related costs have become an important component of the total cost of ownership of this class of systems. Figure 1.8 provides some insight into how energy is used in modern IT equipment by breaking down the peak power usage of one generation of WSCs deployed at Google in 2017 by main component group.

Although this breakdown will vary significantly depending on how systems are configured for a given workload, the graph indicates that CPUs are the dominant energy consumer in WSCs. Interestingly, the first edition of this book showed the relative energy use of the memory system rising to near parity with CPU consumption. Since then that trend has reversed due to a combination of effects. First, sophisticated thermal management has allowed CPUs to run closer to their maximum power envelope, resulting in higher energy consumption per CPU socket. Second, memory technology has shifted away from power hungry FBDIMMs to DDR3 and DDR4 systems with better energy management. Third, DRAM voltage has dropped from 1.8 V down to 1.2 V. Finally, today's systems have a higher ratio of CPU performance per gigabyte of DRAM, possibly as a result of a more challenging technology scaling roadmap for main memory. While increasing bandwidth demands can still reverse these trends, currently, memory power is still significantly smaller than CPU power. It is also worth noting that the power and cooling overhead are relatively small, reflecting generations of improvements in this area, many specific to WSCs. In Chapter 5, we discuss WSC energy efficiency in further detail; see Figure 5.6 for a discussion on power usage *as a function of load.*

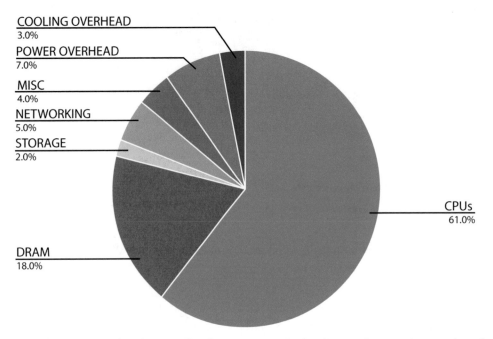

Figure 1.8: Approximate distribution of peak power usage by hardware subsystem in a modern data center using late 2017 generation servers. The figure assumes two-socket x86 servers and 12 DIMMs per server, and an average utilization of 80%.

1.6.6   HANDLING FAILURES AND REPAIRS

The sheer scale of WSCs requires that internet services software tolerates relatively high component fault rates. Disk drives, for example, can exhibit annualized failure rates higher than 4% [PWB07; SG07b]. Different deployments have reported between 1.2 and 16 average server-level restarts per year. With such high component failure rates, an application running across thousands of machines may need to react to failure conditions on an hourly basis. We expand on this topic in Chapter 2, which describes the application domain, and Chapter 7, which deals with fault statistics.

1.7   OVERVIEW OF BOOK

We will elaborate on the issues discussed above in the rest of the book.

Chapter 2 starts with an overview of applications that run on WSCs and that define all the later system design decisions and trade-offs. We discuss key applications like web search and video streaming, and also cover the systems infrastructure stack, including platform-level software, cluster-level infrastructure, and monitoring and management software.

Chapter 3 covers the key hardware building blocks. We discuss the high-level design considerations in WSC hardware and focus on server and accelerator building blocks, storage architectures, and data center networking designs. We also discuss the interplay between compute, storage, and networking, and the importance of system balance.

Chapter 4 looks at the next level of system design, focusing on data center power, cooling infrastructure, and building design. We provide an overview of the basics of the mechanical and electrical engineering involved in the design of WSCs and delve into case studies of how Google designed the power delivery and cooling in some of its data centers.

Chapter 5 discusses the broad topics of energy and power efficiency. We discuss the challenges with measuring energy efficiency consistently and the power usage effectiveness (PUE) metric for data center-level energy efficiency, and the design and benefits from power oversubscription. We discuss the energy efficiency challenges for computing, with specific focus on energy proportional computing and energy efficiency through specialization.

Chapter 6 discusses how to model the total cost of ownership of WSC data centers to address both capital expenditure and operational costs, with case studies of traditional and WSC computers and the trade-offs with utilization and specialization.

Chapter 7 discusses uptime and availability, including data that shows how faults can be categorized and approaches to dealing with failures and optimizing repairs.

Chapter 8 concludes with a discussion of historical trends and a look forward. With the slowing of Moore's Law, we are entering an exciting era of system design, one where WSC data centers and cloud computing will be front and center, and we discuss the various challenges and opportunities ahead.

CHAPTER 2

Workloads and Software Infrastructure

2.1 WAREHOUSE DATA CENTER SYSTEMS STACK

The applications that run on warehouse-scale computers (WSCs) dominate many system design trade-off decisions. This chapter outlines some of the distinguishing characteristics of software that runs in large internet services and the system software and tools needed for a complete computing platform. Here are some terms used to describe the different software layers in a typical WSC deployment.

- *Platform-level software:* The common firmware, kernel, operating system distribution, and libraries expected to be present in all individual servers to abstract the hardware of a single machine and provide a basic machine abstraction layer.

- *Cluster-level infrastructure:* The collection of distributed systems software that manages resources and provides services at the cluster level. Ultimately, we consider these services as an operating system for a data center. Examples are distributed file systems, schedulers and remote procedure call (RPC) libraries, as well as programming models that simplify the usage of resources at the scale of data centers, such as MapReduce [DG08], Dryad [Isa+07], Hadoop [Hadoo], Sawzall [Pik+05], BigTable [Cha+06], Dynamo [DeC+07], Dremel [Mel+10], Spanner [Cor+12], and Chubby [Bur06].

- *Application-level software:* Software that implements a specific service. It is often useful to further divide application-level software into online services and offline computations, since they tend to have different requirements. Examples of online services are Google Search, Gmail, and Google Maps. Offline computations are typically used in large-scale data analysis or as part of the pipeline that generates the data used in online services, for example, building an index of the web or processing satellite images to create map tiles for the online service.

- *Monitoring and development software:* Software that keeps track of system health and availability by monitoring application performance, identifying system bottlenecks, and measuring cluster health.

Figure 2.1 summarizes these layers as part of the overall software stack in WSCs.

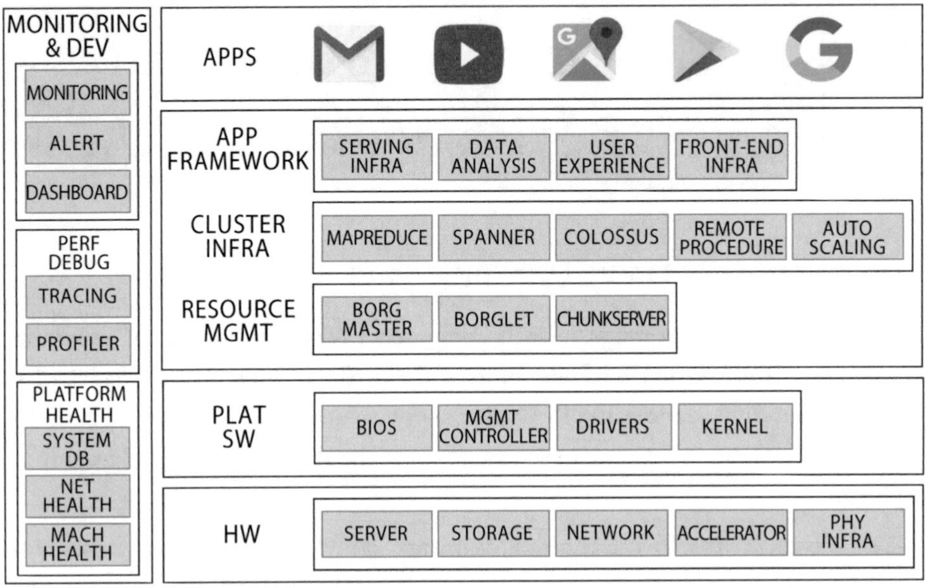

Figure 2.1: Overview of the Google software stack in warehouse-scale computers.

2.2   PLATFORM-LEVEL SOFTWARE

The basic software system image running in WSC server nodes isn't much different than what one would expect on a regular enterprise server platform. Therefore we won't go into detail on this level of the software stack.

Firmware, device drivers, or operating system modules in WSC servers can be simplified to a larger degree than in a general purpose enterprise server. Given the higher degree of homogeneity in the hardware configurations of WSC servers, we can streamline firmware and device driver development and testing since fewer combinations of devices will exist. In addition, a WSC server is deployed in a relatively well known environment, leading to possible optimizations for increased performance. For example, the majority of the networking connections from a WSC server will be to other machines within the same building, and incur lower packet losses than in long-distance internet connections. Thus, we can tune transport or messaging parameters (timeouts, window sizes, and so on) for higher communication efficiency.

Virtualization first became popular for server consolidation in enterprises but now is also popular in WSCs, especially for Infrastructure-as-a-Service (IaaS) cloud offerings [VMware]. A virtual machine provides a concise and portable interface to manage both the security and performance isolation of a customer's application, and allows multiple guest operating systems to co-exist with limited additional complexity. The downside of VMs has always been performance, particularly for I/O-intensive workloads. In many cases today, those overheads are improving and the benefits of the VM model outweigh their costs. The simplicity of VM encapsulation also makes it easier to implement live migration (where a VM is moved to another server without needing to bring down the VM instance). This then allows the hardware or software infrastructure to be upgraded or repaired without impacting a user's computation. Containers are an alternate popular abstraction that allow for isolation across multiple workloads on a single OS instance. Because each container shares the host OS kernel and associated binaries and libraries, they are more lightweight compared to VMs, smaller in size and much faster to start.

2.3   CLUSTER-LEVEL INFRASTRUCTURE SOFTWARE

Much like an operating system layer is needed to manage resources and provide basic services in a single computer, a system composed of thousands of computers, networking, and storage also requires a layer of software that provides analogous functionality at a larger scale. We call this layer the *cluster-level infrastructure*. Three broad groups of infrastructure software make up this layer.

2.3.1   RESOURCE MANAGEMENT

This is perhaps the most indispensable component of the cluster-level infrastructure layer. It controls the mapping of user tasks to hardware resources, enforces priorities and quotas, and provides basic task management services. In its simplest form, it is an interface to manually (and statically) allocate groups of machines to a given user or job. A more useful version would present a higher level of abstraction, automate allocation of resources, and allow resource sharing at a finer level of granularity. Users of such systems would be able to specify their job requirements at a relatively high level (for example, how much CPU performance, memory capacity, and networking bandwidth) and have the scheduler translate those requirements into an appropriate allocation of resources.

Kubernetes (www.kubernetes.io) is a popular open-source program which fills this role that orchestrates these functions for container-based workloads. Based on the ideas behind Google's cluster management system, Borg, Kubernetes provides a family of APIs and controllers that allow users to specify tasks in the popular Open Containers Initiative format (which derives from Docker containers). Several patterns of workloads are offered, from horizontally scaled stateless applications to critical stateful applications like databases. Users define their workloads' resource needs and Kubernetes finds the best machines on which to run them.

It is increasingly important that cluster schedulers also consider power limitations and energy usage optimization when making scheduling decisions, not only to deal with emergencies (such as cooling equipment failures) but also to maximize the usage of the provisioned data center power budget. Chapter 5 provides more detail on this topic, and more information can be found in recent publications [FF05, Lim+13]. Similarly, cluster scheduling must also consider correlated failure domains and fault tolerance when making scheduling decisions. This is discussed further in Chapter 7.

2.3.2 CLUSTER INFRASTRUCTURE

Nearly every large-scale distributed application needs a small set of basic functionalities. Examples are reliable distributed storage, remote procedure calls (RPCs), message passing, and cluster-level synchronization. Implementing this type of functionality correctly with high performance and high availability is complex in large clusters. It is wise to avoid re-implementing such tricky code for each application and instead create modules or services that can be reused. Colossus (successor to GFS) [GGL03, Ser17], Dynamo [DeC+07], and Chubby [Bur06] are examples of reliable storage and lock services developed at Google and Amazon for large clusters.

Many tasks that are amenable to manual processes in a small deployment require a significant amount of infrastructure for efficient operations in large-scale systems. Examples are software image distribution and configuration management, monitoring service performance and quality, and triaging alarms for operators in emergency situations. The Autopilot system from Microsoft [Isa07] offers an example design for some of this functionality for Windows Live data centers. Monitoring the overall health of the hardware fleet also requires careful monitoring, automated diagnostics, and automation of the repairs workflow. Google's System Health Infrastructure, described by Pinheiro et al. [PWB07], is an example of the software infrastructure needed for efficient health management. Finally, performance debugging and optimization in systems of this scale need specialized solutions as well. The X-Trace [Fon+07] system developed at UC Berkeley is an example of monitoring infrastructure aimed at performance debugging of large distributed systems.

2.3.3 APPLICATION FRAMEWORK

The entire infrastructure described in the preceding paragraphs simplifies the deployment and efficient usage of hardware resources, but it does not fundamentally hide the inherent complexity of a large scale system as a target for the average programmer. From a programmer's standpoint, hardware clusters have a deep and complex memory/storage hierarchy, heterogeneous components, failure-prone components, varying adversarial load from other programs in the same system, and resource scarcity (such as DRAM and data center-level networking bandwidth). Some types of higher-level operations or subsets of problems are common enough in large-scale services that it pays off to build targeted programming frameworks that simplify the development of new products. Flume [Cha+10], MapReduce [DG08], Spanner [Cor+12], BigTable [Cha+06], and Dynamo [DeC+07] are good examples of pieces of infrastructure software that greatly improve programmer productivity by automatically handling data partitioning, distribution, and fault tolerance within their respective domains. Equivalents of such software for the cloud, such as Google Kubernetes Engine (GKE), CloudSQL, AppEngine, etc. will be discussed in the discussion about cloud later in this section.

2.4   APPLICATION-LEVEL SOFTWARE

2.4.1   WORKLOAD DIVERSITY

Web search was one of the first large-scale internet services to gain widespread popularity, as the amount of web content exploded in the mid-1990s, and organizing this massive amount of information went beyond what could be accomplished with available human-managed directory services. However, as networking connectivity to homes and businesses continues to improve, offering new services over the internet, sometimes replacing computing capabilities that traditionally lived in the client, has become more attractive. Web-based maps and email services are early examples of these trends.

This increase in the breadth of services offered has resulted in a corresponding diversity in application-level requirements. For example, a search workload may not require an infrastructure capable of high-performance atomic updates and is inherently forgiving of hardware failures (because absolute precision every time is less critical in web search). This is not true for an application that tracks user clicks on sponsored links (ads). Clicks on ads are small financial transactions, which need many of the guarantees expected from a transactional database management system. Figure 2.2 presents the cumulative distribution of cycles across workloads in Google data centers. The top 50 workloads account for only about 60% of the total WSC cycles, with a long tail accounting for the rest of the cycles [Kan+15].

Once we consider the diverse requirements of multiple services, the data center clearly must be a general-purpose computing system. Although specialized hardware solutions might be a good fit for individual parts of services (we discuss accelerators in Chapter 3), the breadth of requirements makes it important to focus on general-purpose system design. Another factor against hardware specialization is the speed of workload churn; product requirements evolve rapidly, and smart programmers will learn from experience and rewrite the baseline algorithms and data structures much more rapidly than hardware itself can evolve. Therefore, there is substantial risk that by the time a specialized hardware solution is implemented, it is no longer a good fit even for the problem area for which it was designed, unless these areas are ones where there is a significant focus on hardware-software codesign. Having said that, there are cases where specialization has yielded significant gains, and we will discuss these further later.

Below we describe workloads used for web search, video serving, machine learning, and citation-based similarity computation. Our objective here is not to describe internet service workloads in detail, especially because the dynamic nature of this market will make those obsolete by publishing time. However, it is useful to describe at a high level a few workloads that exemplify some important characteristics and the distinctions between key categories of online services and batch (offline) processing systems.

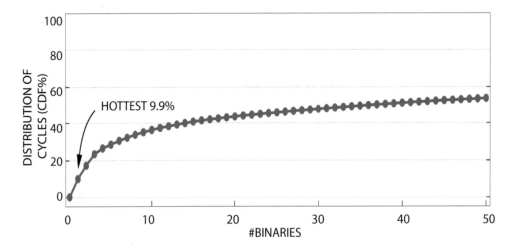

Figure 2.2: Diversity of workloads in WSCs.

2.4.2   WEB SEARCH

This is the quintessential "needle in a haystack" problem. Although it is hard to accurately determine the size of the web at any point in time, it is safe to say that it consists of trillions of individual documents and that it continues to grow. If we assume the web to contain 100 billion documents, with an average document size of 4 KB (after compression), the haystack is about 400 TB. The database for web search is an index built from that repository by inverting that set of documents to create a repository in the logical format shown in Figure 2.3.

A lexicon structure associates an ID to every term in the repository. The *termID* identifies a list of documents in which the term occurs, called a *posting list*, and some contextual information about it, such as position and various other attributes (for example, whether the term is in the document title).

The size of the resulting inverted index depends on the specific implementation, but it tends to be on the same order of magnitude as the original repository. The typical search query consists of a sequence of terms, and the system's task is to find the documents that contain all of the terms (an AND query) and decide which of those documents are most likely to satisfy the user. Queries can optionally contain special operators to indicate alternation (OR operators) or to restrict the search to occurrences of the terms in a particular sequence (phrase operators). For brevity we focus on the more common AND query in the example below.

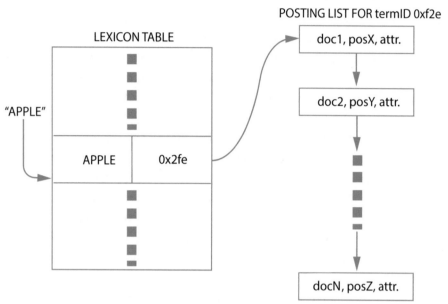

Figure 2.3: Logical view of a web index.

Consider a query such as [*new york restaurants*]. The search algorithm must traverse the posting lists for each term (*new, york, restaurants*) until it finds all documents contained in all three posting lists. At that point it ranks the documents found using a variety of parameters, such as the overall importance of the document (in Google's case, that would be the PageRank score [PB] as well as other properties such as number of occurrences of the terms in the document, positions, and so on) and returns the highest ranked documents to the user.

Given the massive size of the index, this search algorithm may need to run across a few thousand machines. That is accomplished by splitting (or *sharding*) the index into load-balanced subfiles and distributing them across all of the machines. Index partitioning can be done by document or by term. The user query is received by a front-end web server and distributed to all of the machines in the index cluster. As necessary for throughput or fault tolerance, multiple copies of index subfiles can be placed in different machines, in which case only a subset of the machines is involved in a given query. Index-serving machines compute local results, pre-rank them, and send their best results to the front-end system (or some intermediate server), which selects the best results from across the whole cluster. At this point, only the list of doc_IDs corresponding to the resulting web page hits is known. A second phase is needed to compute the actual title, URLs, and a query-specific document snippet that gives the user some context around the search terms. This phase is implemented by sending the list of doc_IDs to a set of machines containing copies of the documents themselves. Once again, a repository this size needs to be partitioned (sharded) and placed in a large number of servers.

The total user-perceived latency for the operations described above needs to be a fraction of a second; therefore, this architecture places heavy emphasis on latency reduction. However, high throughput is also a key performance metric because a popular service may need to support many thousands of queries per second. The index is updated frequently, but in the time granularity of handling a single query, it can be considered a read-only structure. Also, because there is no need for index lookups in different machines to communicate with each other except for the final merge step, the computation is very efficiently parallelized. Finally, further parallelism is available by exploiting the fact that there are no logical interactions across different web search queries.

If the index is sharded by doc_ID, this workload has relatively small networking requirements in terms of average bandwidth because the amount of data exchanged between machines is typically not much larger than the size of the queries themselves (about a hundred bytes or so) but does exhibit some bursty behavior. Basically, the servers at the front-end act as traffic amplifiers as they distribute a single query to a very large number of servers. This creates a burst of traffic not only in the request path but possibly also on the response path as well. Therefore, even if overall network utilization is low, careful management of network flows is needed to minimize congestion.

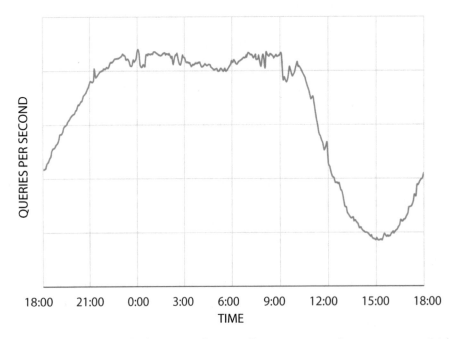

Figure 2.4: Example of daily traffic fluctuation for a search service in one data center over a 24-hr period.

Finally, because web search is an online service, it suffers from normal traffic variations because users are more active on the web at different times of the day. Figure 2.4 illustrates this effect, showing that traffic at peak usage hours can be more than twice as high as off-peak periods. Such variability presents a challenge to system operators because the service must be sized for traffic intensities significantly higher than average.

2.4.3 VIDEO SERVING

IP video traffic represented 73% of the global internet in 2016, and is expected to grow to 83% by 2021. Live video will grow 15-fold, and video-on-demand will grow double by then. In July 2015, users were uploading 400 hr of video per minute to YouTube, and in February 2017 users were watching 1 billion hours of YouTube video per day. Video transcoding (decoding video in one format and encoding it into a different format) is a crucial part of any video sharing infrastructure: video is uploaded in a plethora of combinations of format, codec, resolution, frame rate, color space, and so on. These videos need to be converted to the subset of codecs, resolutions, and formats that client devices can play, and adapted to the network bandwidth available to optimize the user experience.

Video serving has three major cost components: the compute costs due to video transcoding, the storage costs for the video catalog (both originals and transcoded versions), and the network egress costs for sending transcoded videos to end users. Improvements in video compression codecs improve the storage and egress costs at the expense of higher compute costs. The YouTube video processing pipeline balances the three factors based on the popularity profile of videos, only investing additional effort on compressing highly popular videos. Below we describe how video on demand works; other video serving use cases, like on-the-fly transcoding or live video streaming, are similar at a high level but differ in the objective functions being optimized, leading to different architectures.

Every video uploaded to YouTube is first transcoded [Lot+18] from its original upload format to a temporary high-quality common intermediate format, to enable uniform processing in the rest of the pipeline. Then, the video is chunked into segments and transcoded into multiple output resolutions and codecs so that when a user requests the video, the available network bandwidth and the capabilities of the player device are matched with the best version of the video chunk to be streamed. Video chunks are distributed across multiple machines, parallelizing transcoding of each video segment into multiple formats, and optimizing for both throughput and latency. Finally, if a video is identified as highly popular, it will undergo a second video transcoding pass, where additional compute effort is invested to generate a smaller video at the same perceptual quality. This enables users to get a higher resolution version of the video at the same network bandwidth, and the higher compute costs are amortized across egress savings on many playbacks.

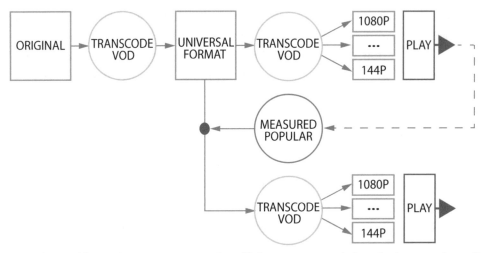

Figure 2.5: The YouTube video processing pipeline. Videos are transcoded multiple times depending on their popularity. VOD = video on demand.

Once video chunks are transcoded to their playback ready formats in the data center, they are distributed through the Edge network, which caches the most recently watched videos to minimize latency and amplify egress bandwidth. When a user requests a video that is not readily available in the Edge caches, the request is forwarded to the nearest data center or a peer Edge location, and the requested video is uploaded from the catalog.

2.4.4   SCHOLARLY ARTICLE SIMILARITY

Services that respond to user requests provide many examples of large-scale computations required for the operation of internet services. These computations are typically data-parallel workloads needed to prepare or package the data that is subsequently used by the online services. For example, computing PageRank or creating inverted index files from a web repository fall in this category. But in this section, we use a different example: finding similar articles in a repository of academic papers and journals. This is a useful feature for internet services that provide access to scientific publications, such as Google Scholar (http://scholar.google.com). Article similarity relationships complement keyword-based search systems as another way to find relevant information; after finding an article of interest, a user can ask the service to display other articles that are strongly related to the original article.

There are several ways to compute similarity scores, and it is often appropriate to use multiple methods and combine the results. With academic articles, various forms of citation analysis are known to provide good quality similarity scores. Here we consider one such type of analysis, called co-citation. The underlying idea is to count every article that cites articles A and B as a vote for the similarity between A and B. After that is done for all articles and appropriately normalized, we obtain a numerical score for the (co-citation) similarity between all pairs of articles and create a data structure that for each article returns an ordered list (by co-citation score) of similar articles. This data structure is periodically updated, and each update then becomes part of the serving state for the online service.

The computation starts with a citation graph that creates a mapping from each article identifier to a set of articles cited by it. The input data is divided into hundreds of files of approximately the same size (for example, by taking a fingerprint of the article identifier, dividing it by the number of input files, and using the remainder as the file ID) to enable efficient parallel execution. We use a sequence of MapReduce runs to take a citation graph and produce a co-citation similarity score vector for all articles. In the first Map phase, we take each citation list (A1, A2, A3, . . . , An) and generate all possible pairs of documents, and then feed them to the Reduce phase, which counts all occurrences of each pair. This first step results in a structure that associates all pairs of co-cited documents with a co-citation count. Note that this becomes much less than a quadratic explosion because most documents have a co-citation count of zero and are therefore omitted. A second MapReduce pass groups all entries for a given document, normalizes their scores, and generates a list of documents with decreasing similarity scores to the original one.

This two-pass data-parallel program executes on hundreds of servers with relatively lightweight computation in each stage followed by significant all-to-all communication between the Map and Reduce workers in each phase. Unlike web search, however, the networking traffic is streaming in nature, which makes it friendlier to existing congestion control algorithms. Also contrary to web search, the latency of individual tasks is much less important than the overall parallel efficiency of the workload.

2.4.5   MACHINE LEARNING

Deep neural networks (DNNs) have led to breakthroughs, such as cutting the error rate in an image recognition competition since 2011 from 26% to 3.5% and beating a human champion at Go [Jou+18]. At Google, DNNs are applied to a wide range of applications including speech, vision, language, translation, search ranking, and many more. Figure 2.6 illustrates the growth of machine learning at Google.

Neural networks (NN) target brain-like functionality and are based on a simple artificial neuron: a nonlinear function (such as max(0, value)) of a weighted sum of the inputs. These artificial neurons are collected into layers, with the outputs of one layer becoming the inputs of the next one in the sequence. The "deep" part of DNN comes from going beyond a few layers, as the large data sets in the cloud allowed more accurate models to be built by using extra and larger layers to capture higher levels of patterns or concepts [Jou+18].

The two phases of DNNs are *training* (or learning) and *inference* (or prediction), and they refer to DNN model development vs. use [Jou+18]. DNN workloads are further be classified into different categories: convolutional, sequence, embedding-based, multilayer perceptron, and reinforcement learning [MLP18].

Training determines the weights or parameters of a DNN, adjusting them repeatedly until the DNN produces the desired results. Virtually all training is in floating point. During training, multiple learners process subsets of the input training set and reconcile the parameters across the learners either through parameter servers or reduction across learners. The learners typically process the input data set through multiple *epochs*. Training can be done asynchronously [Dea+12], with each learner independently communicating with the parameter servers, or synchronously, where learners operate in lockstep to update the parameters after every step. Recent results show that synchronous training provides better model quality; however, the training performance is limited by the slowest learner [Che+16].

Inference uses the DNN model developed during the training phase to make predictions on data. DNN inference is typically user facing and has strict latency constraints [Jou+17]. Inference can be done using floating point (single precision, half precision) or quantized (8-bit, 16-bit) computation. Careful quantization of models trained in floating point is needed to enable inference without any quality loss compared to the floating point models. Lower precision inference enables lower latency and improved power efficiency for inference [Jou+17].

Three kinds of Neural Networks (NNs) are popular today.

1. *Multi-Layer Perceptrons (MLP)*: Each new layer is a set of nonlinear functions of weighted sum of all outputs (*fully connected*) from a prior one.

2. *Convolutional Neural Networks (CNN)*: Each ensuing layer is a set of nonlinear functions of weighted sums of spatially nearby subsets of outputs from the prior layer, which also reuses the weights.

3. *Recurrent Neural Networks (RNN)*: Each subsequent layer is a collection of nonlinear functions of weighted sums of outputs and the previous state. The most popular RNN is *Long Short-Term Memory* (LSTM). The art of the LSTM is in deciding what to forget and what to pass on as state to the next layer. The weights are reused across time steps.

Table 2.1 describes recent versions of six production applications (two examples of each of the three types of NNs) as well as the ResNet50 benchmark [He+15]. One MLP is a recent version of RankBrain [Cla15]; one LSTM is a subset of GNM Translate [Wu+16b]; one CNN is Inception; and the other CNN is DeepMind AlphaGo [Sil+16].

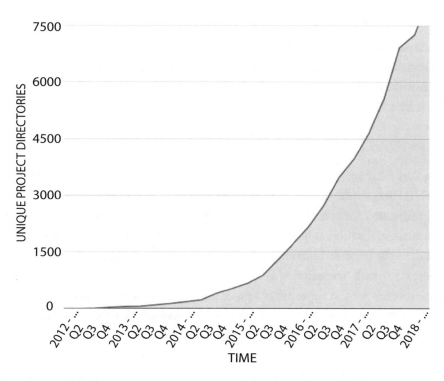

Figure 2.6: Growth of machine learning at Google.

Table 2.1: Six production applications plus ResNet benchmark. The fourth column is the total number of operations (not execution rate) that training takes to converge.

Type of Neural Network	Parameters (MiB)	Training			Inference
		Examples to Convergence	ExaOps to Conv	Ops per Example	Ops per Example
MLP0	225	1 trillion	353	353 Mops	118 Mops
MLP1	40	650 billion	86	133 Mops	44 Mops
LSTM0	498	1.4 billion	42	29 Gops	9.8 Gops
LSTM1	800	656 million	82	126 Gops	42 Gops
CNN0	87	1.64 billion	70	44 Gops	15 Gops
CNN1	104	204 million	7	34 Gops	11 Gops
ResNet	98	114 million	<3	23 Gops	8 Gops

2.5   MONITORING INFRASTRUCTURE

An important part of the cluster-level infrastructure software layer is concerned with various forms of system introspection. Because the size and complexity of both the workloads and the hardware infrastructure make the monitoring framework a fundamental component of any such deployments, we describe it here in more detail.

2.5.1   SERVICE-LEVEL DASHBOARDS

System operators must keep track of how well an internet service is meeting its service level indicators (SLIs). The monitoring information must be very fresh so that an operator (or an automated system) can take corrective actions quickly and avoid significant disruption—within seconds, not minutes. Fortunately, the most critical information needed is restricted to just a few signals that can be collected from the front-end servers, such as latency and throughput statistics for user requests. In its simplest form, such a monitoring system can simply be a script that polls all front-end servers every few seconds for the appropriate signals and displays them to operators in a dashboard. Stackdriver Monitoring [GStaDr] is a publicly available tool which shares the same infrastructure as Google's internal monitoring systems.

Large-scale services often need more sophisticated and scalable monitoring support, as the number of front-ends can be quite large, and more signals are needed to characterize the health of the service. For example, it may be important to collect not only the signals themselves but also their derivatives over time. The system may also need to monitor other business-specific parameters in addition to latency and throughput. The monitoring system supports a simple language that lets operators create derived parameters based on baseline signals being monitored. Finally, the system generates automatic alerts to on-call operators depending on monitored values and thresholds. Fine tuning a system of alerts (or alarms) can be tricky because alarms that trigger too often because of false positives will cause operators to ignore real ones, while alarms that trigger only in extreme cases might get the operator's attention too late to allow smooth resolution of the underlying issues.

2.5.2   PERFORMANCE DEBUGGING TOOLS

Although service-level dashboards help operators quickly identify service-level problems, they typically lack the detailed information required to know *why* a service is slow or otherwise not meeting requirements. Both operators and the service designers need tools to help them understand the complex interactions between many programs, possibly running on hundreds of servers, so they can determine the root cause of performance anomalies and identify bottlenecks. Unlike a service-level dashboard, a performance debugging tool may not need to produce information in real-time for online operation. Think of it as the data center analog of a CPU profiler that determines which function calls are responsible for most of the time spent in a program.

Distributed system tracing tools have been proposed to address this need. These tools attempt to determine all the work done in a distributed system on behalf of a given initiator (such as a user request) and detail the causal or temporal relationships among the various components involved.

These tools tend to fall into two broad categories: black-box monitoring systems and application/middleware instrumentation systems. WAP5 [Rey+06b] and the Sherlock system [Bah+07] are examples of black-box monitoring tools. Their approach consists of observing networking traffic among system components and inferring causal relationships through statistical inference methods. Because they treat all system components (except the networking interfaces) as black boxes, these approaches have the advantage of working with no knowledge of, or assistance from, applications or software infrastructure components. However, this approach inherently sacrifices information accuracy because all relationships must be statistically inferred. Collecting and analyzing more messaging data can improve accuracy but at the expense of higher monitoring overheads.

Instrumentation-based tracing schemes, such as Pip [Rey+06a], Magpie [Bar+03b], and X-Trace [Fon+07], take advantage of the ability to explicitly modify applications or middleware libraries for passing tracing information across machines and across module boundaries within machines. The annotated modules typically also log tracing information to local disks for subsequent collection by an external performance analysis program. These systems can be very accurate as there is no need for inference, but they require all components of the distributed system to be instrumented to collect comprehensive data. The Dapper [Sig+10] system, developed at Google, is an example of an annotation-based tracing tool that remains effectively transparent to application-level software by instrumenting a few key modules that are commonly linked with all applications, such as messaging, control flow, and threading libraries. Stackdriver Trace [GStaDr] is a publicly available implementation of the Dapper system. XRay, a feature of the LLVM compiler, uses compiler instrumentation to add trace points at every function entry and exit, enabling extremely fine latency detail when active and very low overhead when inactive. For even deeper introspection, the Stackdriver Debugger [GStaDb] tool allows users to dynamically add logging to their programs without recompiling or redeploying.

CPU profilers based on sampling of hardware performance counters have been incredibly successful in helping programmers understand microarchitecture performance phenomena. Google-Wide Profiling (GWP) [Ren+10] selects a random subset of machines to collect short whole machine and per-process profile data, and combined with a repository of symbolic information for all Google binaries produces cluster-wide view of profile data. GWP answers questions such as: which is the most frequently executed procedure at Google, or which programs are the largest users of memory? The publicly available Stackdriver Profiler [GStaPr] product is inspired by GWP.

2.5.3   PLATFORM-LEVEL HEALTH MONITORING

Distributed system tracing tools and service-level dashboards measure both health and performance of applications. These tools can infer that a hardware component might be misbehaving, but that is still an indirect assessment. Moreover, because both cluster-level infrastructure and application-level software are designed to tolerate hardware component failures, monitoring at these levels can miss a substantial number of underlying hardware problems, allowing them to build up until software fault-tolerance can no longer mitigate them. At that point, service disruption could be severe. Tools that continuously and directly monitor the health of the computing platform are needed to understand and analyze hardware and system software failures. In Chapter 7, we discuss some of those tools and their use in Google's infrastructure in more detail.

Site reliability engineering

While we have talked about infrastructure software thus far, most WSC deployments support an important function called "site reliability engineering," different from traditional system administration in that software engineers handle day-to-day operational tasks for systems in the fleet. Such SRE software engineers design monitoring and infrastructure software to adjust to load variability and common faults automatically so that humans are not in the loop and frequent incidents are self-healing. An excellent book by a few of Google's site reliability engineers [Mur+16] summarizes additional principles used by SREs in large WSC deployments.

2.6   WSC SOFTWARE TRADEOFFS

2.6.1   DATA CENTER VS. DESKTOP

Software development in internet services differs from the traditional desktop/server model in many ways.

- *Ample parallelism*: Typical internet services exhibit a large amount of parallelism stemming from both data- and request-level parallelism. Usually, the problem is not to find parallelism but to manage and efficiently harness the explicit parallelism inherent in the application. Data parallelism arises from the large data sets of relatively independent records that need processing, such as collections of billions of web pages or billions of log lines. These very large data sets often require significant computation for each parallel (sub) task, which in turn helps hide or tolerate communication and synchronization overheads. Similarly, request-level parallelism stems from the hundreds or thousands of requests per second that popular internet services receive. These requests rarely involve read-write sharing of data or synchronization across requests. For example, search requests are essentially independent and deal with a mostly read-only database; therefore, the computation can be easily partitioned both within a request and across different requests. Similarly, whereas web email transactions do modify user data, requests from different users are essentially independent from each other, creating natural units of data partitioning and concurrency. As long as the update rate is low, even systems with highly interconnected data (such as social networking backends) can benefit from high request parallelism.

- *Workload churn*: Users of internet services are isolated from the services' implementation details by relatively well-defined and stable high-level APIs (such as simple URLs), making it much easier to deploy new software quickly. Key pieces of Google's services have release cycles on the order of a couple of weeks compared to months or years for desktop software products. Google's front-end web server binaries, for example, are released on a weekly cycle, with nearly a thousand independent code changes checked in by hundreds of developers—the core of Google's search services is re-implemented nearly from scratch every 2–3 years. This environment creates significant incentives for rapid product innovation, but makes it hard for a system designer to extract useful benchmarks even from established applications. Moreover, because internet services is still a relatively new field, new products and services frequently emerge, and their success with users directly affects the resulting workload mix in the data center. For example, video services such as YouTube have flourished in relatively short periods and present a very different set of requirements from the existing large customers of computing cycles in the data center, potentially affecting the optimal design point of WSCs. A beneficial side effect of this aggressive software deployment environment is that hardware architects are not necessarily burdened with having to provide good performance for immutable pieces of code. Instead, architects can consider the possibility of significant software rewrites to leverage new hardware capabilities or devices.

- *Platform homogeneity*: The data center is generally a more homogeneous environment than the desktop as a target platform for software development. Large internet services operations typically deploy a small number of hardware and system software configurations at any given time. Significant heterogeneity arises primarily from the incentive to deploy more cost-efficient components that become available over time. Homogeneity within a platform generation simplifies cluster-level scheduling and load balancing and reduces the maintenance burden for platforms software, such as kernels and drivers. Similarly, homogeneity can allow more efficient supply chains and more efficient repair processes because automatic and manual repairs benefit from having more experience with fewer types of systems. In contrast, software for desktop systems can make few assumptions about the hardware or software platform they are deployed on, and their complexity and performance characteristics may suffer from the need to support thousands or even millions of hardware and system software configurations.

- *Fault-free operation*: Because internet service applications run on clusters of thousands of machines—each of them not dramatically more reliable than PC-class hardware— the multiplicative effect of individual failure rates means that some type of fault is expected every few hours or less (more details are provided in Chapter 7). As a result, although it may be reasonable for desktop-class software to assume a fault-free hardware operation for months or years, this is not true for data center-level services: internet services need to work in an environment where faults are part of daily life. Ideally, the cluster-level system software should provide a layer that hides most of that complexity from application-level software, although that goal may be difficult to accomplish for all types of applications.

Although the plentiful thread-level parallelism and a more homogeneous computing platform help reduce software development complexity in internet services compared to desktop systems, the scale, the need to operate under hardware failures, and the speed of workload churn have the opposite effect.

2.6.2   PERFORMANCE AND AVAILABILITY TOOLBOX

Some basic programming concepts tend to occur often in both infrastructure and application levels because of their wide applicability in achieving high performance or high availability in large-scale deployments. The following table (Table 2.2) describes some of the most prevalent concepts. Some articles, by Hamilton [Ham07], Brewer [Bre01], and Vogels [Vog08], provide interesting further reading on how different organizations have reasoned about the general problem of deploying internet services at a very large scale.

Table 2.2: Key concepts in performance and availability trade-offs

Technique	Main Advantages	Description
Replication	Performance and availability	Data replication can improve both throughput and availability. It is particularly powerful when the replicated data is not often modified, since replication makes updates more complex.
Reed-Solomon codes	Availability and space savings	When the primary goal is availability, not throughput, error correcting codes allow recovery from data losses with less space overhead than straight replication.

Sharding (partitioning)	Performance and availability	Sharding splits a data set into smaller fragments (shards) and distributes them across a large number of machines. Operations on the data set are dispatched to some or all of the shards, and the caller coalesces results. The sharding policy can vary depending on space constraints and performance considerations. Using very small shards (or micro-sharding) is particularly beneficial to load balancing and recovery.
Load-balancing	Performance	In large-scale services, service-level performance often depends on the slowest responder out of hundreds or thousands of servers. Reducing response-time variance is therefore critical. In a sharded service, we can load balance by biasing the sharding policy to equalize the amount of work per server. That policy may need to be informed by the expected mix of requests or by the relative speeds of different servers. Even homogeneous machines can offer variable performance characteristics to a load-balancing client if servers run multiple applications. In a replicated service, the load-balancing agent can dynamically adjust the load by selecting which servers to dispatch a new request to. It may still be difficult to approach perfect load balancing because the amount of work required by different types of requests is not always constant or predictable. Microsharding (see above) makes dynamic load balancing easier since smaller units of work can be changed to mitigate hotspots.

Health checking and watchdog timers	Availability	In a large-scale system, failures often manifest as slow or unresponsive behavior from a given server. In this environment, no operation can rely on a given server to make forward progress. Moreover, it is critical to quickly determine that a server is too slow or unreachable and steer new requests away from it. Remote procedure calls must set well-informed timeout values to abort long-running requests, and infrastructure-level software may need to continually check connection-level responsiveness of communicating servers and take appropriate action when needed.
Integrity checks	Availability	In some cases, besides unresponsiveness, faults manifest as data corruption. Although those may be rare, they do occur, often in ways that underlying hardware or software checks do not catch (for example, there are known issues with the error coverage of some networking CRC checks). Extra software checks can mitigate these problems by changing the underlying encoding or adding more powerful redundant integrity checks.
Application-specific compression	Performance	Often, storage comprises a large portion of the equipment costs in modern data centers. For services with very high throughput requirements, it is critical to fit as much of the working set as possible in DRAM; this makes compression techniques very important because the decompression is orders of magnitude faster than a disk seek. Although generic compression algorithms can do quite well, application-level compression schemes that are aware of the data encoding and distribution of values can achieve significantly superior compression factors or better decompression speeds.

Eventual consistency	Performance and availability	Often, keeping multiple replicas up-to-date using the traditional guarantees offered by a database management system significantly increases complexity, hurts performance, and reduces availability of distributed applications [Vog08]. Fortunately, large classes of applications have more relaxed requirements and can tolerate inconsistent views for limited periods, provided that the system eventually returns to a stable consistent state.
Centralized control	Performance	In theory, a distributed system with a single master limits the resulting system availability to the availability of the master. Centralized control is nevertheless much simpler to implement and generally yields more responsive control actions. At Google, we have tended toward centralized control models for much of our software infrastructure (like MapReduce and GFS). Master availability is addressed by designing master failover protocols.
Canaries	Availability	A very rare but realistic catastrophic failure scenario in online services consists of a single request distributed to a very large number of servers, exposing a program-crashing bug and resulting in system-wide outages. A technique often used at Google to avoid such situations is to first send the request to one (or a few) servers and only submit it to the rest of the system upon successful completion of that (canary) request.
Redundant execution and tail-tolerance	Performance	In very large-scale systems, the completion of a parallel task can be held up by the slower execution of a very small percentage of its subtasks. The larger the system, the more likely this situation can arise. Sometimes a small degree of redundant execution of subtasks can result in large speedup improvements.

2.6.3   BUY VS. BUILD

Traditional IT infrastructure makes heavy use of third-party software components, such as databases and system management software, and concentrates on creating software that is specific to the particular business where it adds direct value to the product offering; for example, as business logic on top of application servers and database engines. Large-scale internet service providers such as Google usually take a different approach, in which both application-specific logic and much of the cluster-level infrastructure software is written in-house. Platform-level software does make use of third-party components, but these tend to be open-source code that can be modified in-house as needed. As a result, more of the entire software stack is under the control of the service developer.

This approach adds significant software development and maintenance work but can provide important benefits in flexibility and cost efficiency. Flexibility is important when critical functionality or performance bugs must be addressed, allowing a quick turnaround time for bug fixes at all levels. It eases complex system problems because it provides several options for addressing them. For example, an unwanted networking behavior might be difficult to address at the application level but relatively simple to solve at the RPC library level, or the other way around.

Historically, when we wrote the first edition of this book, a primary reason favoring build versus buy was that the needed warehouse-scale software infrastructure simply was not available commercially. In addition, it is hard for third-party software providers to adequately test and tune their software unless they themselves maintain large clusters. Last, in-house software may be simpler and faster because it can be designed to address only the needs of a small subset of services, and can therefore be made much more efficient in that domain. For example, BigTable omits some of the core features of a traditional SQL database to gain much higher throughput and scalability for its intended use cases, and GFS falls short of offering a fully Posix compliant file system for similar reasons. Today, such models of scalable software development are more widely prevalent. Most of the major cloud providers have equivalent versions of software developed in-house, and open-source software versions of such software are also widely used (e.g., Kubernetes, Hadoop, OpenStack, Mesos).

2.6.4   TAIL-TOLERANCE

Earlier in this chapter we described a number of techniques commonly used in large-scale software systems to achieve high performance and availability. As systems scale up to support more powerful online web services, we have found that such techniques are insufficient to deliver service-wide responsiveness with acceptable tail latency levels. (*Tail latency* refers to the latency of the slowest requests, that is, the tail of the latency distribution.) Dean and Barroso [DB13] have argued that at large enough scale, simply stamping out all possible sources of performance variability in individual system components is as impractical as making all components in a large fault-tolerant system fault-free.

Consider a hypothetical system where each server typically responds in 10 ms but with a 99th percentile latency of 1 s. In other words, if a user request is handled on just one such server, 1 user request in 100 will be slow (take 1 s). Figure 2.7 shows how service level latency in this hypothetical scenario is impacted by very modest fractions of latency outliers as cluster sizes increase. If a user request must collect responses from 100 such servers in parallel, then 63% of user requests will take more than 1 s (marked as an "x" in the figure). Even for services with only 1 in 10,000 requests experiencing over 1 s latencies at the single server level, a service with 2,000 such servers will see almost 1 in 5 user requests taking over 1 s (marked as an "o") in the figure.

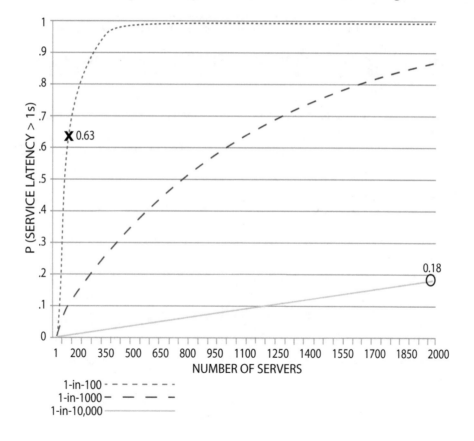

Figure 2.7: Probability of >1 s service level response time as the system scale and frequency of server level high latency outliers varies.

Dean and Barroso show examples of programming techniques that can tolerate these kinds of latency variability and still deliver low tail latency at the service level. The techniques they propose often take advantage of resource replication that has already been provisioned for fault-tolerance, thereby achieving small additional overheads for existing systems. They predict that tail tolerant techniques will become more invaluable in the next decade as we build ever more formidable online web services.

2.6.5   LATENCY NUMBERS THAT ENGINEERS SHOULD KNOW

This section is inspired by Jeff Dean's summary of key latency numbers that engineers should know [Dea09]. These rough operation latencies help engineers reason about throughput, latency, and capacity within a first-order approximation. We have updated the numbers here to reflect technology and hardware changes in WSC.

Table 2.3: Latency numbers that every WSC engineer should know. (Updated version of table from [Dea09].)

Operation	Time
L1 cache reference	1.5 ns
L2 cache reference	5 ns
Branch misprediction	6 ns
Uncontended mutex lock/unlock	20 ns
L3 cache reference	25 ns
Main memory reference	100 ns
Decompress 1 KB with Snappy [Sna]	500 ns
"Far memory"/Fast NVM reference	1,000 ns (1us)
Compress 1 KB with Snappy [Sna]	2,000 ns (2us)
Read 1 MB sequentially from memory	12,000 ns (12 us)
SSD Random Read	100,000 ns (100 us)
Read 1 MB bytes sequentially from SSD	500,000 ns (500 us)
Read 1 MB sequentially from 10Gbps network	1,000,000 ns (1 ms)
Read 1 MB sequentially from disk	10,000,000 ns (10 ms)
Disk seek	10,000,000 ns (10 ms)
Send packet California→Netherlands→California	150,000,000 ns (150 ms)

2.7   CLOUD COMPUTING

Recently, cloud computing has emerged as an important model for replacing traditional enterprise computing systems with one that is layered on top of WSCs. The proliferation of high speed internet makes it possible for many applications to move from the traditional model of on-premise computers and desktops to the cloud, running remotely in a commercial provider's data center. Cloud computing provides efficiency, flexibility, and cost saving. The cost efficiency of cloud computing is achieved through co-locating multiple virtual machines on the same physical hosts to increase utilization. At a high level, a virtual machine (VM) is similar to other online web services, built on top of cluster-level software to leverage the entire warehouse data center stack. Although a VM-based workload model is the simplest way to migrate on-premise computing to WSCs, it comes with some additional challenges: I/O virtualization overheads, availability model, and resource isolation. We discuss these in more detail below.

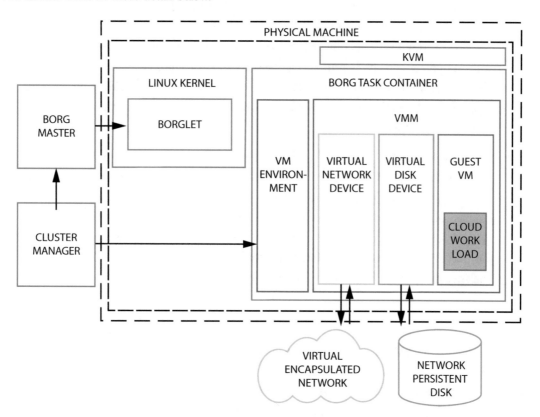

Figure 2.8: Overview of VM-based software stack for Google Cloud Platform workloads.

- *I/O Virtualization*: A VM does not have direct access to hardware resources like local hard drives or networking. All I/O requests go through an abstraction layer, such as *virtio*, in the guest operating system. The hypervisor or virtual machine monitor (VMM) translates the I/O requests into the appropriate operations: For example, storage requests are redirected to the network persistent disk or local SSD drives, and networking requests are sent through virtualized network for encapsulation. Although I/O virtualization often incurs some performance overhead improvements in both virtualization techniques and hardware support for virtualization have steadily reduced these overheads [Dal+18].

- *Availability model*: Large-scale distributed services achieve high availability by running multiple instances of a program within a data center, and at the same time maintaining N + 1 redundancy at the data center level to minimize the impact of scheduled maintenance events. However, many enterprise applications have fewer users and use older stacks, such as relational databases, where horizontal scaling is often not possible. Cloud providers instead use *live migration* technology to ensure high availability by moving running VMs out of the way of planned maintenance events, including system updates and configurations changes. As a recent example, when responding to the Meltdown and Spectre vulnerabilities [Hen+18]. Google performed host kernel upgrades and security patches to address these issues across its entire fleet without losing a single VM.

Resource isolation: As discussed earlier, the variability in latency of individual components is amplified at scale at the service level due to interference effects. Resource quality of service (QoS) has proved to be an effective technique for mitigating the impact of such interference. In cloud computing, malicious VMs can exploit multi-tenant features to cause severe contention on shared resources, conducting Denial of Service (DoS) and side-channel attacks. This makes it particularly important to balance the tradeoff between security guarantees and resource sharing.

2.7.1 WSC FOR PUBLIC CLOUD SERVICES VS. INTERNAL WORKLOADS

Google began designing WSCs nearly a decade before it began offering VM-based public clouds as an enterprise product, the Google Cloud Platform [GCP]. During that time Google's computing needs were dominated by search and ads workloads, and therefore it was natural consider the design of both WSCs as well as its corresponding software stack in light of the requirements of those specific workloads. Software and hardware engineers took advantage of that relatively narrow set of requirements to vertically optimize their designs with somewhat tailored, in-house solutions across the whole stack and those optimizations resulted in a company-internal development environment that diverged from that of early public cloud offerings. As the portfolio of internal workloads broadened, Google's internal designs had to adapt to consider more general purpose use cases and as a consequence, the gap between internal workload requirements and those of public clouds has narrowed significantly in most application areas. Although some differences still remain, internal and external developers experience a much more similar underlying platform today, which makes it possible to bring internal innovations to external users much more quickly than would have been possible a decade ago. One example is the recent public availability of TPUs to Google Cloud Platform customers, discussed more in Chapter 3.

2.7.2   CLOUD NATIVE SOFTWARE

The rising popularity of Cloud Services brought with it a new focus on software that takes full advantage of what cloud providers have to offer. Coupled with the shift toward containers as the vehicle for workloads, the "Cloud Native" ethos emphasizes properties of clouds that are not easily realized in private data centers. This includes things like highly dynamic environments, API-driven self-service operations, and instantaneous, on-demand resources. Clouds are elastic in a way that physical WSCs have difficulty matching. These properties allow developers to build software that emphasizes scalability and automation, and minimizes operational complexity and toil.

Just as containers and orchestrators catalyzed this shift, other technologies are frequently adopted at the same time. Microservices are the decomposition of larger, often monolithic, applications into smaller, limited-purpose applications that cooperate via strongly defined APIs, but can be managed, versioned, tested, and scaled independently. Service meshes allow application operators to decouple management of the application itself from management of the networking that surrounds it. Service discovery systems allow applications and microservices to find each other in volatile environments, in real time.

2.8   INFORMATION SECURITY AT WAREHOUSE SCALE

Cloud users depend on the ability of the WSC provider to be a responsible steward of user or customer data. Hence, most providers invest considerable resources into a sophisticated and layered security posture. The at-scale reality is that security issues will arise, mandating a holistic approach that considers prevention, detection, and remediation of issues. Many commercial software or hardware vendors offer individual point solutions designed to address specific security concerns, but few (if any) offer solutions that address end-to-end security concerns for WSC. The scope is as broad as the scale is large, and serious WSC infrastructure requires serious information security expertise.

At the hardware level, there are concerns about the security of the physical data center premises. WSC providers employ technologies such as biometric identification, metal detection, cameras, vehicle barriers, and laser-based intrusion detection systems, while also constraining the number of personnel even permitted onto the data center floor (for example, see https://cloud.google.com/security/infrastructure/design/). Effort is also invested to ascertain the provenance of hardware and the designs on which it is based. Some WSC providers invest in custom solutions for cryptographic machine identity and the security of the initial firmware boot processes. Examples include Google's Titan, Microsoft's Cerberus, and Amazon's Nitro.

Services deployed on this infrastructure use encryption for inter-service communication; establish systems for service identity, integrity, and isolation; implement access control mechanisms on top of identity primitives; and ultimately carefully control and audit access to end user or customer data. Even with the correct and intended hardware, issues can arise. The recent Spectre/Meltdown/L1TF issues [Hen+18] are illustrative of the need for WSC providers to maintain defense-in-depth and expertise to assess risk and develop and deploy mitigations. WSC providers employ coordinated management systems that can deploy CPU microcode, system software updates, or other patches in a rapid but controlled manner. Some even support live migration of customer VMs to allow the underlying machines to be upgraded without disrupting customer workload execution.

Individual challenges, such as that of authenticating end user identity, give rise to a rich set of security and privacy challenges on their own. Authenticating users in the face of password reuse and poor password practices at scale, and otherwise preventing login or account recovery abuse often require dedicated teams of experts.

Data at rest must always be encrypted, often with both service- and device-centric systems with independent keys and administrative domains. Storage services face a form of cognitive dissonance, balancing durability requirements against similar requirements to be able to truly, confidently assert that data has been deleted or destroyed. These services also place heavy demands on the network infrastructure in many cases, often requiring additional computational resources to ensure the network traffic is suitably encrypted.

Internal network security concerns must be managed alongside suitable infrastructure to protect internet-scale communication. The same reliability constraints that give rise to workload migration infrastructure also require network security protocols that accommodate the model of workloads disconnected from their underlying machine. One example is Google application layer transport security (ALTS) [Gha+17]. The increasing ubiquity of SSL/TLS, and the need to respond appropriately for ever-present DoS concerns, require dedicated planning and engineering for cryptographic computation and traffic management.

Realizing the full benefit of best-practices infrastructure security also requires operational security effort. Intrusion detection, insider risk, securing employee devices and credentials, and even basic things such as establishing and adhering to best practices for safe software development all require non-trivial investment.

As information security begins to truly enjoy a role as a first-class citizen in WSC infrastructure, some historical industry norms are being reevaluated. For example, it may be challenging for a WSC to entrust critical portions of its infrastructure to an opaque or black-box third-party component. The ability to return machines to a known state—all the way down to the firmware level—in between potentially mutually distrusting workloads is becoming a must-have.

CHAPTER 3

WSC Hardware Building Blocks

As mentioned earlier, the architecture of WSCs is largely defined by the hardware building blocks chosen. This process is analogous to choosing logic elements for implementing a microprocessor, or selecting the right set of chipsets and components for a server platform. In this case, the main building blocks are server hardware, networking fabric, and storage hierarchy components. This chapter focuses on these building blocks, with the objective of increasing the intuition needed for making such choices.

3.1 SERVER HARDWARE

Clusters of mid-range servers are the preferred building blocks for WSCs today [BDH03]. This is true for a number of reasons, the primary one being the underlying cost-efficiency of mid-range servers when compared with the high-end shared memory systems that had earlier been the preferred building blocks for the high-performance and technical computing space. The continuing CPU core count increase has also reached a point that most VM/task instances can comfortably fit into a two-socket server. Such server platforms share many key components with the high-volume personal computing market, and therefore benefit more substantially from economies of scale. It is typically hard to do meaningful cost-efficiency comparisons because prices fluctuate and performance is subject to benchmark characteristics and the level of effort put into benchmarking. In the first edition, we showed a comparison of TPC-C benchmark [TPC] results from a system based on a high-end server (HP Integrity Superdome-Itanium2 [TPC07a]) and one based on a low-end server (HP ProLiant ML350 G5 [TPC07b]). The difference in cost-efficiency was over a factor of four in favor of the low-end server. When looking for more recent data for this edition, we realized that there are no competitive benchmarking entries that represent the high-end server design space. As we observed in 2009, the economics of the server space made that class of machines occupy a small niche in the marketplace today. The more interesting discussions now are between mid-range server nodes and extremely low end (so-called "wimpy") servers, which we cover later in this chapter.

3.1.1 SERVER AND RACK OVERVIEW

Servers hosted in individual racks are the basic building blocks of WSCs. They are interconnected by hierarchies of networks, and supported by the shared power and cooling infrastructure.

As discussed in Chapter 1, WSCs use a relatively homogeneous hardware and system software platform. This simplicity implies that each server generation needs to provide optimized performance and cost for a wide range of WSC workloads and the flexibility to support their resource requirements. Servers are usually built in a tray or blade enclosure format, housing the motherboard, chipset, and additional plug-in components. The motherboard provides sockets and plug-in slots to install CPUs, memory modules (DIMMs), local storage (such as Flash SSDs or HDDs), and network interface cards (NICs) to satisfy the range of resource requirements. Driven by workload performance, total cost of ownership (TCO), and flexibility, several key design considerations determine the server's form factor and functionalities.

- *CPU*: CPU power, often quantified by the thermal design power, or TDP; number of CPU sockets and NUMA topology; CPU selection (for example, core count, core and uncore frequency, cache sizes, and number of inter-socket coherency links).

- *Memory*: Number of memory channels, number of DIMMs per channel, and DIMM types supported (such as RDIMM, LRDIMM, and so on).

- *Plug-in IO cards*: Number of PCIe cards needed for SSD, NIC, and accelerators; form factors; PCIe bandwidth and power, and so on.

- *Tray-level power and cooling, and device management and security options*: Voltage regulators, cooling options (liquid versus air-cooled), board management controller (BMC), root-of-trust security, and so on.

- *Mechanical design*: Beyond the individual components, how they are assembled is also an important consideration: server form-factors (width, height, depth) as well as front or rear access for serviceability.

Figure 3.1 shows a high-level block diagram of the key components of a server tray. Figure 3.2 shows photographs of server trays using Intel and IBM processors.

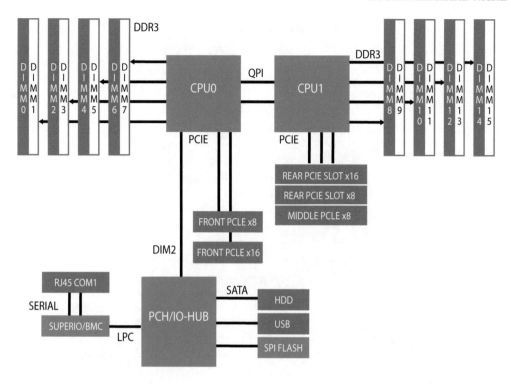

Figure 3.1: Block diagram of a server.

Going into more system details, the first x86-server supports two Intel Haswell CPU sockets, and one Wellsburg Platform Controller Hub (PCH). Each CPU can support up to 145W TDP (for example, Intel's 22 nm-based Haswell processor with 18-core per socket and 45MB L3 shared cache). The server has 16 DIMM slots, supporting up to two DIMMs per memory channel (2DPC) with ECC DRAM. With Integrated Voltage Regulators, the platforms allow per core DVFS. The system supports 80 PCIe Gen3 lanes (40 lanes per CPU), and various PCIe plug-ins with different IO width, power, and form factors, allowing it to host PCIe cards for SSD, 40 GbE NIC, accelerators. It also includes several SATA ports and supports both direct-attached storage and PCIe-attached storage appliance ("disk trays").

Figure 3.2: Example server trays: (top) Intel Haswell-based server tray and (bottom) IBM Power8-based server tray.

The second server is similar, except it supports IBM Power CPUs with higher thread counts and TDP, more PCIe lanes (96 lanes per two-socket), and up to 32 DDR3 DIMMs (twice that of the two-socket Intel-based system). The system maximizes the memory and IO bandwidth supported by the platform, in order to support a wide range of workloads and accelerators. We also balance the choice of CPU, in terms of core count, frequency, and cache sizes, with the available memory system bandwidth.

The rack is the physical structure that holds tens of servers together. Racks not only provide the physical support structures, but also handle shared power infrastructure, including power delivery, battery backup, and power conversion (such as AC to 48V DC). The width and depth of racks vary across WSCs: some are classic 19-in wide, 48-in deep racks, while others can be wider or shallower. The width and depth of a rack can also constrain server form factor designs. It is often convenient to connect the network cables at the top of the rack, such a rack-level switch is appropriately called a Top of Rack (TOR) switch. These switches are often built using merchant switch silicon, and further interconnected into scalable network topologies (such as Clos) described in more detail below. For example, Google's Jupiter network uses TOR switches with 64x 40 Gbps ports. These ports are split between downlinks that connect servers to the TOR, and uplinks that connect the TOR to the rest of the WSC network fabric. The ratio between number of downlinks and uplinks is called the *oversubscription ratio*, as it determines how much the intra-rack fabric is over-provisioned with respect to the data center fabric.

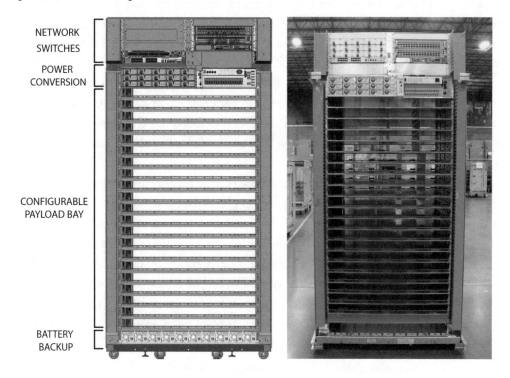

Figure 3.3: Machine racks like this support servers, storage, and networking equipment in Google's data centers.

Figure 3.3 shows an example rack in a Google data center. Aside from providing configurable physical structure for server trays with different widths and heights (for example, it can host four Haswell servers per row), it also provides the configurable power conversion and battery backup support, with redundancy, to match the IT power load. The TOR switch and rack management unit are located at the top of the rack and are further connected to a data center fabric to connect many racks.

The Open Compute Project (http://opencompute.org) also contains detailed specifications of many hardware components for WSCs.

3.1.2   THE IMPACT OF LARGE SMP COMMUNICATION EFFICIENCY

Simple processor-centric cost-efficiency analyses do not account for the fact that large Shared-Memory Multiprocessors (SMPs) benefit from drastically superior intercommunication performance than clusters of low-end servers connected by commodity fabrics. Nodes in a large SMP may communicate at latencies on the order of 100 ns, whereas LAN-based networks, usually deployed in clusters of servers, will experience latencies at or above 100 μs. For parallel applications that fit within a single large SMP (for example, SAP HANA), the efficient communication can translate into dramatic performance gains. However, WSC workloads are unlikely to fit within an SMP, therefore it is important to understand the relative performance of clusters of large SMPs with respect to clusters of low-end servers each with smaller number of CPU sockets or cores (albeit the same server-class CPU cores as used in large SMPs). The following simple model can help make these comparisons.

Assume that a given parallel task execution time can be roughly modeled as a fixed local computation time plus the latency penalty of accesses to global data structures. If the computation fits into a single large shared memory system, those global data accesses will be performed at roughly DRAM speeds (~100 ns). If the computation fits in only a multiple of such nodes, some global accesses will be much slower, on the order of typical LAN speeds (~100 μs). Assume further that accesses to the global store are uniformly distributed among all nodes, so that the fraction of global accesses that map to the local node is inversely proportional to the number of nodes in the system. If the fixed local computation time is of the order of 1 ms—a reasonable value for high-throughput internet services—the equation that determines the program execution time is as follows:

Execution time = 1 ms + f * [100 ns/# nodes + 100 μs * (1 – 1/# nodes)],

where the variable *f* is the *number of global accesses per work unit* (1 ms). In Figure 3.4, we plot the execution time of this parallel workload as the number of nodes involved in the computation increases. Three curves are shown for different values of f, representing workloads with light communication (*f* = 1), medium communication (*f* = 10), and high communication (*f* = 100) patterns. Note that in our model, the larger the number of nodes, the higher the fraction of remote global accesses.

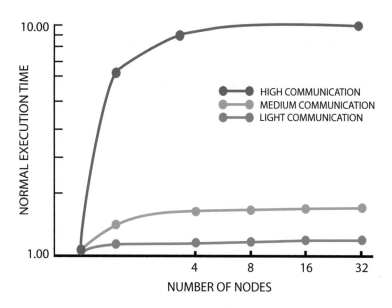

Figure 3.4: Execution time of parallel tasks as the number of SMP nodes increases for three levels of communication intensity. Execution time is normalized to the single node case and plotted in logarithmic scale.

The curves in Figure 3.4 have two interesting aspects worth highlighting. First, under light communication, there is relatively small performance degradation from using clusters of multiple nodes. For medium- and high-communication patterns, the penalties can be quite severe, but they are most dramatic when moving from a single node to two, with rapidly decreasing additional penalties for increasing the cluster size. Using this model, the performance advantage of a single 128-processor SMP over a cluster of thirty-two 4-processor SMPs could be more than a factor of 10×.

By definition, WSC systems will consist of thousands of processor cores. Therefore, we would like to use this model to compare the performance of a cluster built with large SMP servers with one built with low-end ones. Here we assume that the per-core performance is the same for both systems and that servers are interconnected using an Ethernet-class fabric. Although our model is exceedingly simple (for example, it does not account for contention effects), it clearly captures the effects we are interested in.

In Figure 3.5, we apply our model to clusters varying between 512 and 4,192 cores and show the performance advantage of an implementation using large SMP servers (128 cores in a single shared memory domain) versus one using low-end servers (four-core SMPs) The figure compares the performance of clusters with high-end SMP systems to low-end systems, each having between 512 and 4,192 cores, over three communication patterns.. Note how quickly the performance edge of the cluster based on high-end servers deteriorates as the cluster size increases. If the application requires more than 2,000 cores, a cluster of 512 low-end servers performs within approximately 5% of one built with 16 high-end servers, even under a heavy communication pattern. With a performance gap this low, the price premium of the high-end server (4–20 times higher) renders it an unattractive option.

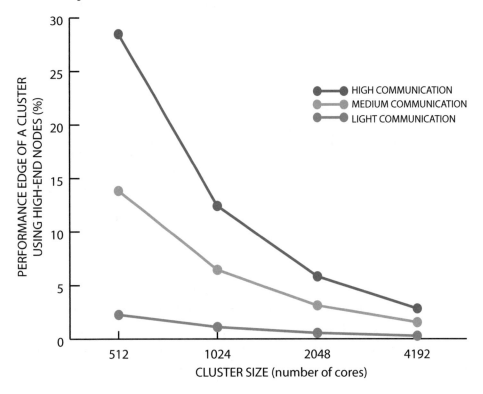

Figure 3.5: Deteriorating performance advantage of a cluster built with large SMP server nodes (128-core SMP) over a cluster with the same number of processor cores built with low-end server nodes (four-core SMP), for clusters of varying size.

The point of this analysis is qualitative in nature: it primarily illustrates how we need to reason differently about our baseline platform choice when architecting systems for applications that are too large for any single high-end server. The broad point is that the performance effects that matter most are those that benefit the system *at the warehouse scale*. Performance enhancements that have the greatest impact on computation, that are local to a single node (such as fast SMP-style communication in our example), are still very important. But if they carry a heavy additional cost, their cost-efficiency may not be as competitive for WSCs as they are for small-scale computers.

Since the first edition of this book was released, the number of cores per processor has steadily increased, allowing larger cluster sizes with smaller numbers of individual server nodes, and has essentially moved us toward the left side of Figure 3.5. However, given the scale of warehouse-scale systems, the discussion above still holds qualitatively. The increasing number of cores per single system does have implications on the broader system balance, and this is discussed further in the last section of this chapter. This analysis framework can also be useful in reasoning about intra-rack and data center-level networking bandwidth provisioning.

3.1.3   BRAWNY VS. WIMPY SERVERS

Clearly one could use the argument laid out above to go further and use CPU cores that are even smaller or wimpier than today's server-class CPU cores. The Piranha chip multiprocessor [Bar+00] was one of the earliest systems to advocate the use of lower-end cores in enterprise-class server systems. In [BDH03], we argued that chip multiprocessors using this approach are especially compelling for Google workloads. More recently, even more radical approaches that leverage embedded-class CPUs (wimpy processors) have been proposed as possible alternatives for WSC systems. Lim et al. [Lim+08], for example, make the case for exactly such alternatives as being advantageous to low-end server platforms once all power-related costs are considered (including amortization of data center build-out costs and the cost of energy). Hamilton [Ham09] makes a similar argument, although using PC-class components instead of embedded ones. The advantages of using smaller, slower CPUs are very similar to the arguments for using mid-range commodity servers instead of high-end SMPs.

- Multicore CPUs in mid-range servers typically carry a price-performance premium over lower-end processors so that the same amount of throughput can be bought two to five times cheaper with multiple smaller CPUs.

- Many applications are memory- or I/O-bound so that faster CPUs do not scale well for large applications, further enhancing the price advantage of simpler CPUs.

- Slower CPUs tend to be more power efficient; typically, CPU power decreases by $O(k^2)$ when CPU frequency decreases by a factor of k.

The FAWN (Fast Array of Wimpy Nodes) project [And+11] at Carnegie Mellon has explored the utility of wimpy cores as the basis for building an energy efficient key-value storage system, with an emphasis on flash memory. The nature of the workload in FAWN storage servers makes them a good fit for less powerful cores since the computation tends to be more I/O- and memory latency-bound than CPU-bound.

Several commercial products have also explored designs with a small number of mobile-class cores. Such systems often provide an integrated interconnect to attach disks, flash storage, and Ethernet ports that can be shared among the servers. For example, HP's Moonshot Servers [HPM13] defined a blade-style chassis that can accommodate 45 server cartridges, including mobile x86 or ARM-based CPUs in a 4.3U form factor. More recently, Microsoft's Project Olympus [MPO] discussed using ARM-based CPUs for search, storage, and machine learning workloads.

In the previous edition of this book, we summarized some of the tradeoffs with very low-performing cores that can make them unattractive for WSCs (a point discussed at length by one of the authors [Höl10]. Specifically, although many internet services benefit from seemingly unbounded request- and data-level parallelism, such systems are not immune from Amdahl's law. As the number of offered parallel threads increases, it can become increasingly difficult to reduce serialization and communication overheads, limiting either speedup or scaleup [DWG92, Lim+08]. In the limit, the amount of inherently serial work performed on behalf of a user request by extremely slow single-threaded hardware will dominate overall execution time.

Also, the more the number of threads that handle a parallelized request, the larger the *variability* in response times from all these parallel tasks, exacerbating the tail latency problem discussed in Chapter 2. One source of large performance variability that occurs on multi-core architectures is from opportunistic overclocking, with vendor-specific names such as Turbo Boost or Turbo CORE. The premise behind this feature is to run the CPU at higher frequencies when there are sufficient electrical and thermal margins to do so. The largest beneficiaries of this feature are single-threaded sequential workloads, which can receive up to 76% higher CPU frequency than the nominal processor frequency.[3] This level of performance variability has several effects: it degrades Amdahl's law by inflating single-threaded performance and further exacerbates variability in response times for distributed scale-out applications by adding a complex performance dimension (number of active CPU cores, electrical, and thermal margins on the CPU). As multi-core processors continue to scale up core counts, addressing this source of heterogeneity within the system becomes a more pressing concern.

As a result, although hardware costs may diminish, software development costs may increase because more applications must be explicitly parallelized or further optimized. For example, suppose that a web service currently runs with a latency of 1-s per user request, half of it caused by CPU time. If we switch to a cluster with lower-end servers whose single-thread performance is three times slower, the service's response time will double to 2-s and application developers may have to spend a substantial amount of effort to optimize the code to get back to the 1-s latency level.

Networking requirements also increase with larger numbers of smaller systems, increasing networking delays and the cost of networking (since there are now more ports in an already expensive switching fabric). It is possible to mitigate this effect by locally interconnecting a small number of slower servers to share a network link, but the cost of this interconnect may offset some of the price advantage gained by switching to cheaper CPUs.

---

[3] https://ark.intel.com/products/codename/37572/Skylake#@server

Smaller servers may also lead to lower utilization. Consider the task of allocating a set of applications across a pool of servers as a bin packing problem—each of the servers is a bin, with as many applications as possible packed inside each bin. Clearly, that task is harder when the bins are small because many applications may not completely fill a server and yet use too much of its CPU or RAM to allow a second application to coexist on the same server.

Finally, even embarrassingly parallel algorithms are sometimes intrinsically less efficient when computation and data are partitioned into smaller pieces. That happens, for example, when the stop criterion for a parallel computation is based on global information. To avoid expensive global communication and global lock contention, local tasks may use heuristics based on local progress only, and such heuristics are naturally more conservative. As a result, local subtasks may execute longer than they might if there were better hints about global progress. Naturally, when these computations are partitioned into smaller pieces, this overhead tends to increase.

A study by Lim et al. [Lim+13] illustrates some of the possible perils of using wimpy cores in a WSC workload. The authors consider the energy efficiency of Atom-based (wimpy) and Xeon-based (brawny) servers while running a memcached server workload. While the Atom CPU uses significantly less power than the Xeon CPU, a cluster provisioned with Xeon servers outperforms one provisioned with Atom servers by a factor of 4 at the same power budget.

Also, from the perspective of cloud applications, most workloads emphasize single-VM performance. Additionally, a bigger system is more amenable to being deployed and sold as smaller VM shapes, but the converse is not true. For these reasons, cloud prefers brawny systems as well.

However, in the past few years there have been several developments that make this discussion more nuanced, with more CPU options *between* conventional wimpy cores and brawny cores. For example, several ARMv8-based servers have emerged with improved performance (for example, Cavium ThunderX2 [CTX2] and Qualcomm Centriq 2400 [QC240]), while Intel Xeon D processors use brawny cores in low-power systems on a chip (SoCs). These options allow server builders to choose from a range of wimpy and brawny cores that best fit their requirements. As a recent example, Facebook's Yosemite microserver uses the one-socket Xeon D CPU for its scale-out workloads and leverages the high-IPC cores to ensure low-latency for web serving. To reduce the network costs, the design shares one NIC among four SoC server cards.

As a rule of thumb, a lower-end server building block must have a healthy cost-efficiency advantage over a higher-end alternative to be competitive. At the moment, the sweet spot for many large-scale services seems to be at the low-end range of server-class machines. We expect more options to populate the spectrum between wimpy and brawny cores, and WSC server design to continue to evolve with these design options.

3.2   COMPUTING ACCELERATORS

Historically, the deployment of specialized computing accelerators (non-general-purpose CPUs) in WSCs has been very limited. As discussed in the second edition of this book, although they promised greater computing efficiencies, such benefits came at the cost of drastically restricting the number of workloads that could benefit from them. However, this has changed recently. Traditional improvements from Moore's law scaling of general-purpose systems has been slowing down. But perhaps more importantly, deep learning models began to appear and be widely adopted, enabling specialized hardware to power a broad spectrum of machine learning solutions. WSC designs responded to these trends. For example, Google not only began to more widely deploy GPUs, but also initiated a program to build further specialized computing accelerators for deep learning algorithms [Jou+17]. Similarly, Microsoft initiated at program to deploy FPGA-based accelerators in their fleet [Put+14].

Neural network (NN) workloads (described in Chapter 2) execute extremely high numbers of floating point operations. Figure 3.6, from OpenAI, shows the growth of compute requirements for neural networks [OAI18]. Since 2013, AI training compute requirements have doubled every 3.5 months. The growth of general-purpose compute has significantly slowed down, with a doubling rate now exceeding 4 or more years (vs. 18–24 months expected from Moore's Law). To satisfy the growing compute needs for deep learning, WSCs deploy GPUs and other specialized accelerator hardware.

Project Catapult (Microsoft) is the most widely deployed example of using reconfigurable accelerator hardware to support DNNs. They chose FPGAs over GPUs to reduce power as well as the risk that latency-sensitive applications wouldn't map well to GPUs. Google not only began to widely deploy GPUs but also started a program to build specialized computing accelerators as the Tensor Processing Units (TPU) [Jou+17]. The TPU project at Google began with FPGAs, but we abandoned them when we saw that FPGAs at that time were not competitive in performance compared to GPUs, and TPUs could use much less power than GPUs while being as fast or faster, giving them potentially significant benefits over both FPGAs and GPUs.

Time to convergence is a critical metric for ML training. Faster time to convergence improves ML model development due to faster training iterations that enable efficient model architecture and hyperparameter exploration. As described inChapter 2, multiple learners or replicas are typically used to process input examples. Increasing the number of learners, however, can have a detrimental impact on model accuracy, depending upon the model and mode of training (synchronous vs. asynchronous). For deep learning inference, many applications are user-facing and have strict response latency deadlines.

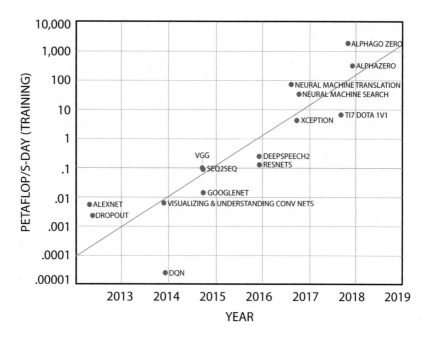

Figure 3.6: Training compute requirements for models over time [OAI18].

3.2.1   GPUs

GPUs are configured with a CPU host connected to a PCIe-attached accelerator tray with multiple GPUs. GPUs within the tray are connected using high-bandwidth interconnects such as NVlink. Multiple GPU trays are connected to the data center network with NICs. The GPUs and NICs communicate directly through PCIe without data transfer through the host. Training on GPUs can be performed synchronously or asynchronously, with synchronous training providing higher model accuracy. Synchronous training has two phases in the critical path: a compute phase and a communication phase that reconciles the parameters across learners. The performance of such a synchronous system is limited by the slowest learner and slowest messages through the network. Since the communication phase is in the critical path, a high performance network that can enable fast reconciliation of parameters across learners with well-controlled tail latencies is important for high-performance deep learning training. Figure 3.7 shows a network-connected pod of GPUs used for training.

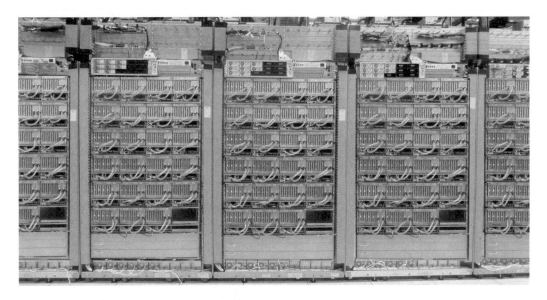

Figure 3.7: Interconnected GPUs for training.

3.2.2   TPUS

While well suited to ML workloads, GPUs still are relatively general purpose devices, and in recent years designers have further specialized them to ML-specific ASICs that drop any vestiges of graphics or high-precision functional units. TPUs are used for training and inference. TPUv1 is an inference-focused accelerator connected to the host CPU through PCIe links; a detailed architecture and performance review can be found in [Jou+17].

TPUv2, in contrast, is a very different ASIC focused on training workloads (Figure 3.8). Each TPU board is connected to one dual socket server. Inputs for training are fed to the system using the data center network from storage racks. Figure 3.8 also shows the block diagram of each TPUv2 chip. Each TPUv2 consists of two Tensor cores. Each Tensor core has a systolic array for matrix computations (MXU) and a connection to high bandwidth memory (HBM) to store parameters and intermediate values during computation.

Multiple TPUv2 accelerator boards are connected through a custom high bandwidth torus network (Figure 3.9) to provide 11 petaflops of ML compute. The accelerator boards in the TPUv2 *pod* work in lockstep to train a deep learning model using synchronous training [Dea]. The high bandwidth network enables fast parameter reconciliation with well-controlled tail latencies, allowing near ideal scalability for training across a pod [Dea].

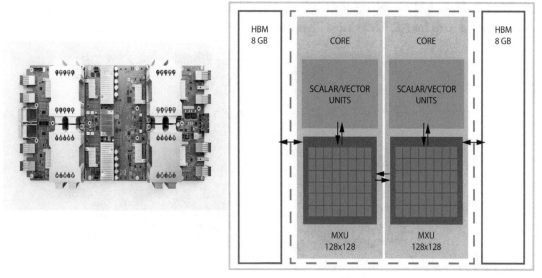

Figure 3.8: TPUv2.

Figure 3.9: Four-rack TPUv2 pod.

TPUv3 is the first liquid-cooled accelerator in Google's data center. Liquid cooling enables TPUv3 to provide eight times the ML compute of TPUv2, with the TPUv3 pod providing more than 100 petaflops of ML compute. Such supercomputing-class computational power supports dramatic new capabilities. For example, AutoML [GCAML], coupled with the computing power of TPUs, enables rapid neural architecture search and faster advances in ML research. Figure 3.10 shows a board with four TPUv3 chips. Figure 3.11 shows a pod of third-generation TPUs.

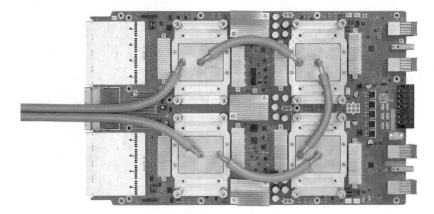

Figure 3.10: TPUv3.

Figure 3.11: Eight-rack TPUv3 pod.

3.3   NETWORKING

3.3.1   CLUSTER NETWORKING

Servers must be connected, and as the performance of servers increases over time, the demand for inter-server bandwidth naturally increases as well. But while we can double the aggregate compute capacity or the aggregate storage simply by doubling the number of compute or storage elements, networking has no straightforward horizontal scaling solution. Doubling leaf bandwidth is easy; with twice as many servers, we'll have twice as many network ports and thus twice as much bandwidth. But if we assume that every server needs to talk to every other server, we need to double not just leaf bandwidth but *bisection bandwidth*, the bandwidth across the narrowest line that equally divides the cluster into two parts. (Using bisection bandwidth to characterize network capacity is common since randomly communicating processors must send about half the bits across the "middle" of the network.)

Unfortunately, doubling bisection bandwidth is difficult because we can't just buy (or make) an arbitrarily large switch. Switch chips are pin- and power-limited in size; for example, a typical merchant silicon switch chip can support a bisection bandwidth of about 1 Tbps (16x 40 Gbps ports) and no chips are available that can do 10 Tbps. We can build larger switches by cascading these switch chips, typically in the form of a fat tree or Clos network,[4] as shown in Figure 3.12 [Mys+09, Vah+10].

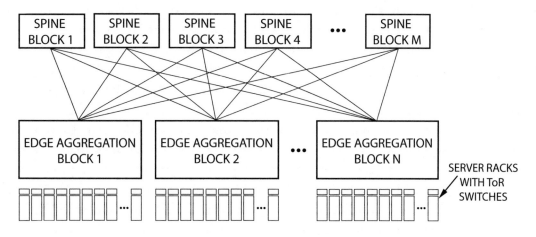

Figure 3.12: Sample three-stage fat tree topology. With appropriate scheduling this tree can deliver the same throughput as a single-stage crossbar switch.

Such a tree using k-port switches can support full throughput among $k^3/4$ servers using $5k^2/4$ switches, allowing networks with tens of thousands of ports. However, the cost of doing so increases significantly because each path to another server now involves more ports. In the simplest network (a single stage consisting of a single central switch), each path consists of two ports: switch in and switch out. The above three-stage network quintuples that to 10 ports, significantly increasing costs. So as bisection bandwidth grows, the cost per connected server grows as well. Port costs can be substantial, especially if a link spans more than a few meters, thus requiring an optical interface. Today the optical components of a 100 m 10 Gbps link can easily cost several hundred

---

[4]   Clos networks are named after Charles Clos, who first formalized their properties in 1952.

dollars (including the cost of the two optical ports, fiber cable, fiber termination, and installation), not including the networking components themselves (switches and NICs).

To reduce costs per machine, WSC designers often oversubscribe the network at the top-of-rack switch. Generally speaking, a rack contains a small enough number of servers so they can be connected with a switch at a reasonable cost. It's not hard to find switches with 48 ports to inter-connect 48 servers at full speed (say, 40 Gbps). In a full fat tree, each such switch would need the same number of ports facing "upward" into the cluster fabric: the edge switches in the figure above devote half their ports to connecting servers, and half to the fabric. All those upward-facing links in turn require more links in the aggregation and core layers, leading to an expensive network. In an oversubscribed network, we increase that 1:1 ratio between server and fabric ports. For example, with 2:1 oversubscription, we build a fat tree for only half the bandwidth, reducing the size of the tree and thus its cost, but also reducing the available bandwidth per server by a factor of two. Each server can still peak at 40 Gbps of traffic, but if all servers are simultaneously sending traffic, they'll only be able to average 20 Gbps. In practice, oversubscription ratios of 4–10 are common. For example, a 48-port switch could connect 40 servers to 8 uplinks, for a 5:1 oversubscription (8 Gbps per server).

Another way to tackle network scalability is to offload some traffic to a special-purpose net-work. For example, if storage traffic is a big component of overall traffic, we could build a separate network to connect servers to storage units. If that traffic is more localized (not all servers need to be attached to all storage units) we could build smaller-scale networks, thus reducing costs. His-torically, that's how all storage was networked: a SAN (storage area network) connected servers to disks, typically using FibreChannel (FC [AI11]) networks rather than Ethernet. Today, Ethernet is becoming more common since it offers comparable speeds, and protocols such as FibreChannel over Ethernet) (FCoE [AI462]), SCSI over IP (iSCSI [iSC03]), and more recently NVMe over Fabric (NVMeoF [NVM]) allow Ethernet networks to integrate well with traditional SANs.

Figure 3.13 shows the structure of Google's Jupiter Clos network [Sin+15]. This multi-stage network fabric uses low-radix switches built from merchant silicon, each supporting 16x 40 Gbps ports. Each 40 G port could be configured in 4x10 G or 40 G mode. A server is connected to its ToR switch using 40 Gbps Ethernet NICs. Jupiter's primary building block is Centauri, a 4RU chassis housing two line cards, each with two switch chips. In an example ToR configuration, each switch chip is configured with 48x10 G to servers and 16x10 G to the fabric, yielding an oversub-scription ratio of 3:1. Servers can also be configured with 40 G mode to have 40 G burst bandwidth.

The ToR switches are connected to layers of aggregation blocks to increase the scale of the network fabric. Each Middle Block (MB) has four Centauri chassis. The logical topology of an MB is a two-stage blocking network, with 256x10 G links available for ToR connectivity and 64x40 G available for connectivity to the rest of the fabric through the spine blocks.

Jupiter uses the same Centauri chassis as the building block for the aggregation blocks. Each ToR chip connects to eight middle blocks with dual redundant 10G links. This aids fast reconver-

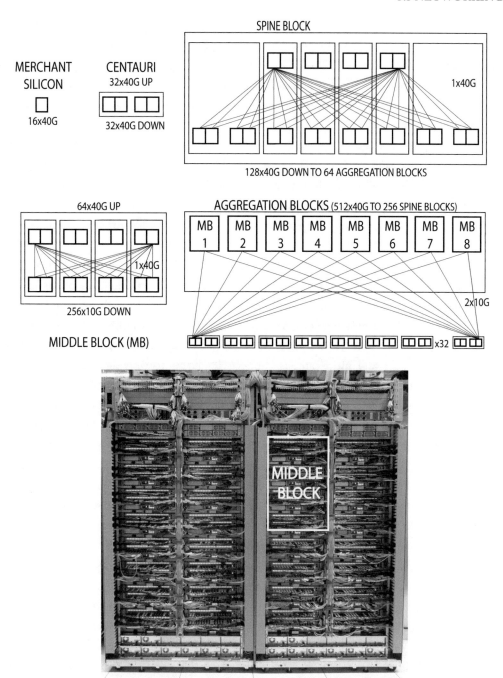

Figure 3.13: Google's Jupiter network. Starting from NIC, to ToR, to multiple switches, to campus networks. Top: (1) switch chip and Centauri chassis, (2) middle block, and (3) spine block and topology. Bottom: A Jupiter rack with middle block highlighted.

gence for the common case of single link failure or maintenance. Each aggregation block exposes 512x40 G (full pop) or 256x40 G (depop) links toward the spine blocks. Six Centauri chassis are grouped in a spine block exposing 128x40 G ports toward the aggregation blocks. Jupiter limits the size to 64 aggregation blocks for dual redundant links between each spine block and aggregation block pair at the largest scale, once again for local reconvergence on single link failure. At this maximum size, the bisection bandwidth is 1.3 petabits per second.

Jupiter employs a separate aggregation block for external connectivity, which provides the entire pool of external bandwidth to each aggregation block. As a rule of thumb, 10% of aggregate intra-cluster bandwidth is allocated for external connectivity using one to three aggregation blocks. These aggregation blocks are physically and topologically identical to those used for ToR connectivity. However, the ports normally employed for ToR connectivity are reallocated to connect to external fabrics.

With Jupiter, the intra-cluster fabric connects directly to the inter-cluster networking layer with Cluster Border Routers (CBRs). Multiple clusters are deployed within the same building and multiple buildings on the same campus. The job scheduling and resource allocation infrastructure leverages campus-level and building-level locality. The design further replaces vendor-based inter cluster switching with Freedome, a two-stage fabric that uses the Border Gateway Protocol (BGP) at both the inter-cluster and intra-campus connectivity layers to provide massive inter-cluster bandwidth within buildings and the campus at a lower cost than existing solutions.

Compared to WSCs, High-Performance Computing (HPC) supercomputer clusters often have a much lower ratio of computation to network bandwidth, because applications such as weather simulations distribute their data across RAM in all nodes, and nodes need to update neighboring nodes after performing relatively few floating-point computations. As a result, traditional HPC systems have used proprietary interconnects with leading-edge link bandwidths, much lower latencies (especially for common functions like barrier synchronizations or scatter/gather operations, which often are directly supported by the interconnect), and some form of a global address space (where the network is integrated with CPU caches and virtual addresses). Typically, such interconnects offer throughputs that are at least an order of magnitude higher than contemporary Ethernet or InfiniBand solutions, but are much more expensive.

WSCs using VMs (or, more generally, task migration) pose further challenges to networks since connection endpoints (that is, IP address/port combinations) can move from one physical machine to another. Typical networking hardware as well as network management software don't anticipate such moves and in fact often explicitly assume that they're not possible. For example, network designs often assume that all machines in a given rack have IP addresses in a common subnet, which simplifies administration and minimizes the number of required forwarding table entries routing tables. More importantly, frequent migration makes it impossible to manage the network manually; programming network elements needs to be automated, so the same cluster manager that decides the placement of computations also needs to update the network state.

The need for a programmable network has led to much interest in OpenFlow (http://www.openflow.org/), P4 (www.p4.org), and software-defined networking (SDN), which move the network control plane out of individual switches into a logically centralized controller [Höl12, ONF12, Jai+13, Sin+15]. Controlling a network from a logically centralized server offers many advantages; in particular, common networking algorithms such as computing reachability, shortest paths, or max-flow traffic placement become much simpler to solve compared to their implementation in current networks where each individual router must solve the same problem while dealing with limited visibility (direct neighbors only), inconsistent network state (routers that are out of sync with the current network state), and many independent and concurrent actors (routers). Network management operations also become simple because a global view can be used to move a network domain, often consisting of thousands of individual switches, from one consistent state to another while simultaneously accounting for errors that may require rollback of the higher-level management operation underway. Recent developments in P4 further enable a protocol- and switch-independent high-level language that allows for programming of packet-forwarding data planes, further increasing flexibility.

In addition, servers are easier to program, offering richer programming environments and much more powerful hardware. As of 2018, a typical router control processor consists of a Xeon-based 2-4 core SoC with two memory channels and 16 GB DRAM. The centralization of the control plane into a few servers also makes it easier to update their software. SDN is a natural match for data center networking, since the applications running in a WSC are already managed by a central entity, the cluster manager. Thus it is natural for the cluster manager to also configure any network elements that applications depend on. SDN is equally attractive to manage WAN networks, where logically centralized control simplifies many routing and traffic engineering problems [Höl12, Jai+13].

For more details on cluster networking, see these excellent recent overview papers: Singh et al. [Sin+15], Kumar et al. [Kum+15], Vahdat et al. [Vah+10], Abts and Felderman [AF12], and Abts and Kim [AK11].

3.3.2   HOST NETWORKING

On the host networking side, WSC presents unique challenges and requirements: high throughputs and low latency with efficient use of host CPU, low tail latencies, traffic shaping (pacing and rate limiting), OS-bypass, stringent security and line-rate encryption, debuggability, QoS and congestion control, etc. Public cloud computing further requires features such as support for virtualization and VM migration. Two general approaches can be combined to meet these requirements: onload where the host networking software leverages host CPU to provide low latency and rich features, and offload that uses compute in NIC cards for functions such as packet processing and crypto. There has been active research and development in this area, for example Azure's use of FPGA as bump-in-the-wire [Fir+18], Amazon's customized NIC for bare-metal support, Google's host-side traffic shaping [Sae+17], and Andromeda approach toward cloud network virtualization [Dal+18].

3.4   STORAGE

The data manipulated by WSC workloads tends to fall into two categories: data that is private to individual running tasks and data that is part of the shared state of the distributed workload. Private data tends to reside in local DRAM or disk, is rarely replicated, and its management is simplified by virtue of its single user semantics. In contrast, shared data must be much more durable and is accessed by a large number of clients, thus requiring a much more sophisticated distributed storage system. We discuss the main features of these WSC storage systems next.

3.4.1   DISK TRAYS AND DISKLESS SERVERS

Figure 3.14 shows an example of a disk tray used at Google that hosts tens of hard drives (22 drives in this case) and provides storage over Ethernet for servers in the WSC. The disk tray provides power, management, mechanical, and network support for these hard drives, and runs a customized software stack that manages its local storage and responds to client requests over RPC.

In traditional servers, local hard drives provide direct-attached storage and serve as the boot/logging/scratch space. Given that most of the traditional needs from the storage device are now handled by the network attached disk trays, servers typically use one local (and a much smaller) hard drive as the boot/logging device. Often, even this disk is removed (perhaps in favor of a small flash device) to avoid the local drive from becoming a performance bottleneck, especially with an increasing number of CPU cores/threads, leading to diskless servers.

A recent white paper [Bre+16] provides more details on the requirements of hard drives for WSCs, focusing on the tradeoffs in the design of disks, tail latency, and security. While current hard drives are designed for enterprise servers, and not specifically for WSC use-case, this paper argues that such "cloud disks" should aim at a global optimal in view of five key metrics: (1) higher I/Os per second (IOPS), typically limited by seeks; (2) higher capacity; (3) lower tail latency when used in WSCs; (4) meeting security requirements; and (5) lower total cost of ownership (TCO). The shift in use case and requirements also creates new opportunities for hardware vendors to explore new physical design and firmware optimizations.

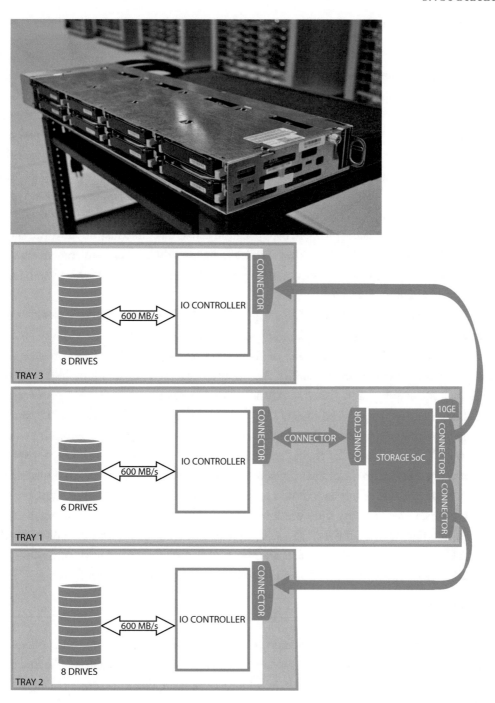

Figure 3.14: (Top) Photograph of a disk tray. (Bottom) Block diagram.

3.4.2   UNSTRUCTURED WSC STORAGE

Google's GFS [GGL03] is an example of a storage system with a simple file-like abstraction (Google's Colossus system has since replaced GFS, but follows a similar architectural philosophy so we choose to describe the better known GFS here). GFS was designed to support the web search indexing system (the system that turned crawled web pages into index files for use in web search), and therefore focuses on high throughput for thousands of concurrent readers/writers and robust performance under high hardware failures rates. GFS users typically manipulate large quantities of data, and thus GFS is further optimized for large operations. The system architecture consists of a primary server (master), which handles metadata operations, and thousands of chunkserver (secondary) processes running on every server with a disk drive, to manage the data chunks on those drives. In GFS, fault tolerance is provided by replication across machines instead of within them, as is the case in RAID systems. Cross-machine replication allows the system to tolerate machine and network failures and enables fast recovery, since replicas for a given disk or machine can be spread across thousands of other machines.

Although the initial version of GFS supported only simple replication, Colossus and its externally available cousin GCS have added support for more space-efficient Reed-Solomon codes, which tend to reduce the space overhead of replication by roughly a factor of two over simple replication for the same level of availability. An important factor in maintaining high availability is distributing file chunks across the whole cluster in such a way that a small number of correlated failures is extremely unlikely to lead to data loss. Colossus optimizes for known possible correlated fault scenarios and attempts to distribute replicas in a way that avoids their co-location in a single fault domain. Wide distribution of chunks across disks over a whole cluster is also key for speeding up recovery. Since replicas of chunks in a given disk are spread across possibly all machines in a storage cluster, reconstruction of lost data chunks is performed in parallel at high speed. Quick recovery is important since long recovery time windows leave under-replicated chunks vulnerable to data loss, in case additional faults hit the cluster. A comprehensive study of availability in distributed file systems at Google can be found in Ford et al. [For+10]. A good discussion of the evolution of file system design at Google can also be found in McKusik and Quinlan [McKQ09].

3.4.3   STRUCTURED WSC STORAGE

The simple file abstraction of Colossus and GCS may suffice for systems that manipulate large blobs of data, but application developers also need the WSC equivalent of database-like functionality, where data sets can be structured and indexed for easy small updates or complex queries. Structured distributed storage systems, such as Google's Bigtable [Cha+06] and Amazon's DynamoDB [DeC+07], were designed to fulfill those needs. Compared to traditional database systems, Bigtable and DynamoDB sacrifice some features, such as the richness of schema representation and strong consistency, in favor of higher performance and availability at massive scales. Bigtable, for example, presents a simple multi-dimensional sorted map consisting of row keys (strings) associated with multiple values organized in columns, forming a distributed sparse table space. Column values are associated with timestamps in order to support versioning and time-series.

The choice of eventual consistency in Bigtable and DynamoDB shifts the burden of resolving temporary inconsistencies to the applications using these systems. A number of application developers within Google have found it inconvenient to deal with weak consistency models and the limitations of the simple data schemes in Bigtable. Second-generation structured storage systems such as Megastore [Bak+11] and subsequently Spanner [Cor+12] were designed to address such concerns. Both Megastore and Spanner provide richer schemas and SQL-like functionality while providing simpler, stronger consistency models. Megastore sacrifices write throughput in order to provide synchronous replication. Spanner uses a new time base API to efficiently serialize globally-distributed transactions, providing a simpler consistency model to applications that need seamless wide-area replication for fault tolerance. Both Megastore and Spanner sacrifice some efficiency in order to provide a simpler programming interface.

Clickstream and Ads data management, for example, is an important use-case of structured storage systems. Such systems require high availability, high scalability of NoSQL systems, and the consistency and usability of SQL databases. Google's F1 system [Shu+13] uses Spanner as datastore, and manages all AdWords data with database features such as distributed SQL queries, transactionally consistent secondary indexes, asynchronous schema changes, optimistic transactions, and automatic change history recording and publishing. The Photon [Ana+13] scalable streaming system supports joining multiple continuously flowing streams of data in real-time with high scalability and low latency, with exactly-once semantics (to avoid double-charging or missed clicks) eventually.

At the other end of the structured storage spectrum from Spanner are systems that aim almost exclusively at high performance. Such systems tend to lack support for transactions or geographic replication, use simple key-value data models, and may have loose durability guarantees. Memcached [Fit+03], developed as a distributed DRAM-based object caching layer, is a popular example at the simplest end of the spectrum. The Stanford RAMCloud [Ous+09] system also uses a distributed DRAM-based data store but aims at much higher performance (over one million lookup operations per second per server) as well as durability in the presence of storage node failures. The FAWN-KV [And+11] system also presents a key-value high-performance storage system but instead uses NAND flash as the storage medium, and has an additional emphasis on energy efficiency, a subject we cover more extensively in Chapter 5.

3.4.4   INTERPLAY OF STORAGE AND NETWORKING TECHNOLOGY

The success of WSC distributed storage systems can be partially attributed to the evolution of data center networking fabrics. Ananthanarayanan et al. [Ana+11] observe that the gap between networking and disk performance has widened to the point that disk locality is no longer relevant in intra-data center computations. This observation enables dramatic simplifications in the design of distributed disk-based storage systems as well as utilization improvements, since any disk byte in a WSC facility can, in principle, be utilized by any task regardless of their relative locality.

Flash devices pose a new challenge for data center networking fabrics. A single enterprise flash device can achieve well over 100x the operations throughput of a disk drive, and one server machine with multiple flash SSDs could easily saturate a single 40 Gb/s network port even within a rack. Such performance levels will stretch not only data center fabric bisection bandwidth but also require more CPU resources in storage nodes to process storage operations at such high rates. Looking ahead, rapid improvements in WSC network bandwidth and latency will likely match flash SSD performance and reduce the importance of flash locality. However, emerging non-volatile memory (NVM) has the potential to provide even higher bandwidth and sub-microsecond access latency. Such high-performance characteristics will further bridge the gap between today's DRAM and flash SSDs, but at the same time present an even bigger challenge for WSC networking.

3.5   BALANCED DESIGNS

Computer architects are trained to solve the problem of finding the right combination of performance and capacity from the various building blocks that make up a WSC. In this chapter we discussed many examples of how the right building blocks are apparent only when one considers the entire WSC system. The issue of balance must also be addressed at this level. It is important to characterize the kinds of workloads that will execute on the system with respect to their consumption of various resources, while keeping in mind three important considerations.

- Smart programmers may be able to restructure their algorithms to better match a more inexpensive design alternative. There is opportunity here to find solutions by software-hardware co-design, while being careful not to arrive at machines that are too complex to program.

- The most cost-efficient and balanced configuration for the hardware may be a match with the combined resource requirements of multiple workloads and not necessarily a perfect fit for any one workload. For example, an application that is seek-limited may not fully use the capacity of a very large disk drive but could share that space with an application that needs space mostly for archival purposes.

- Fungible resources tend to be more efficiently used. Provided there is a reasonable amount of connectivity within a WSC, effort should be put on creating software systems that can flexibly utilize resources in remote servers. This affects balanced system design decisions in many ways. For instance, effective use of remote disk drives may require that the networking bandwidth to a server be equal or higher to the combined peak bandwidth of all the disk drives locally connected to the server.

The right design point depends on more than the high-level structure of the workload itself because data size and service popularity also play an important role. For example, a service with huge data sets but relatively small request traffic may be able to serve most of its content directly from disk drives, where storage is cheap (in dollars per GB) but throughput is low. Very popular services that either have small data set sizes or significant data locality can benefit from in-memory serving instead.

Finally, workload churn in this space is also a challenge to WSC architects. It is possible that the software base may evolve so fast that a server design choice becomes suboptimal during its lifetime (typically three to four years). This issue is even more important for the WSC as a whole because the lifetime of a data center facility generally spans several server lifetimes, or more than a decade or so. In those cases it is useful to try to envision the kinds of machinery or facility upgrades that may be necessary over the lifetime of the WSC system and take that into account during the design phase of the facility.

3.5.1   SYSTEM BALANCE: STORAGE HIERARCHY

Figure 3.15 shows a programmer's view of storage hierarchy of a hypothetical WSC. As discussed earlier, the server consists of a number of processor sockets, each with a multicore CPU and its internal cache hierarchy, local shared and coherent DRAM, a number of directly attached disk drives, and/or flash-based solid state drives. The DRAM and disk/flash resources within the rack are accessible through the first-level rack switches (assuming some sort of remote procedure call API to them exists), and all resources in all racks are accessible via the cluster-level switch. The relative balance of various resources depends on the needs of target applications. The following configuration assumes an order of magnitude less flash capacity than traditional spinning media since that is roughly the relative cost per byte difference between these two technologies.

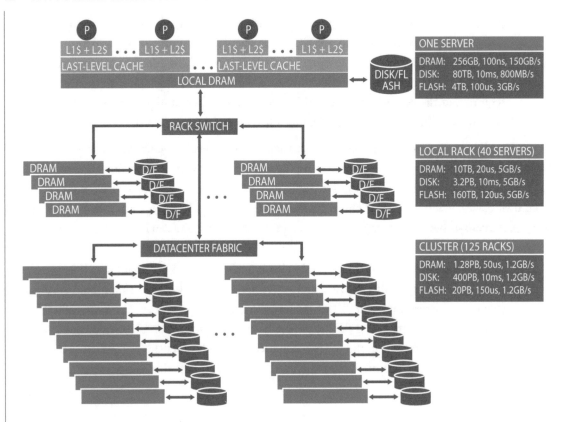

Figure 3.15: Storage hierarchy of a WSC.

3.5.2   QUANTIFYING LATENCY, BANDWIDTH, AND CAPACITY

Figure 3.16 attempts to quantify the latency, bandwidth, and capacity characteristics of a WSC. For illustration we assume a system with 5,000 servers, each with 256 GB of DRAM, one 4 TB SSD, and eight 10 TB disk drives. Each group of 40 servers is connected through a 40-Gbps link to a rack-level switch that has an additional 10-Gbps uplink bandwidth per machine for connecting the rack to the cluster-level switch (an oversubscription factor of four). Network latency numbers assume a TCP/IP transport, and networking bandwidth values assume that each server behind an oversubscribed set of uplinks is using its fair share of the available cluster-level bandwidth. For disks, we show typical commodity disk drive (SATA) latencies and transfer rates.

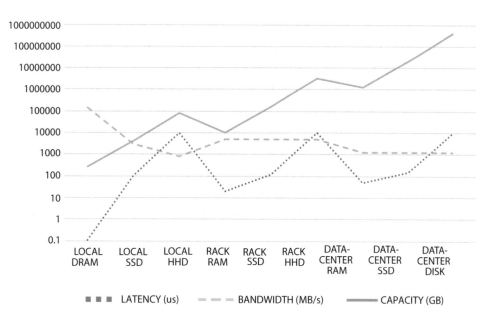

Figure 3.16: Latency, bandwidth, and capacity of WSC storage hierarchy levels.

The graph shows the relative latency, bandwidth, and capacity of each resource pool. For example, the bandwidth available from local SSDs is about 3 GB/s, whereas the bandwidth from off-rack SSDs is just 1.25 GB/s via the shared rack uplinks. On the other hand, total disk storage in the cluster is more than one million times larger than local DRAM.

A large application that requires many more servers than can fit on a single rack must deal effectively with these large discrepancies in latency, bandwidth, and capacity. These discrepancies are much larger than those seen on a single machine, making it more difficult to program a WSC. A key challenge for architects of WSCs is to smooth out these discrepancies in a cost-efficient manner. Conversely, a key challenge for software architects is to build cluster infrastructure and services that hide most of this complexity from application developers. For example, NAND flash technology, originally developed for portable electronics, has found target use cases in WSC systems. Flash-based SSDs are a viable option for bridging the cost and performance gap between DRAM and disks, as displayed in Figure 3.16. Flash's most appealing characteristic with respect to disks is its performance under random read operations, which is nearly three orders of magnitude better. In fact, flash's performance is so high that it becomes a challenge to use it effectively in distributed storage systems since it demands much higher bandwidth from the WSC fabric, as well as microsecond performance support from the hardware/software stack. Note that in the worst case, writes to flash can be several orders of magnitude slower than reads, and garbage collection can further increase write amplification and tail latency. Characteristics of flash around read/write asymmetry, read/write interference, and garbage collection behaviors introduce new challenges and opportunities in adopting low-latency storage tiers in a balanced WSC design.

Emerging technologies such as non-volatile memories (NVM) (for example, Intel 3D Xpoint based memory [3DX] and fast SSD products such as Samsung Z-NAND [Sam17]) add another tier between today's DRAM and flash/storage hierarchy. NVM has the potential to provide cheaper and more scalable alternatives to DRAM, which is fast approaching its scaling bottleneck, but also presents challenges for WSC architects who now have to consider data placement, prefetching, and migration over multiple memory/storage tiers. NVM and flash also present new performance and efficiency challenges and opportunities, as traditional system design and software optimizations lack support for their microsecond (μs)-scale latencies. A new set of hardware and software technologies are needed to provide a simple programming model to achieve high performance [Bar+17].

CHAPTER   4

Data Center Basics: Building, Power, and Cooling

Internet and cloud services run on a planet-scale computer with workloads distributed across multiple data center buildings around the world. These data centers are designed to house computing, storage, and networking infrastructure. The main function of the buildings is to deliver the utilities needed by equipment and personnel there: power, cooling, shelter, and security. By classic definitions, there is little work produced at the data center. Other than some departing photons, all of the energy consumed is converted into heat. The delivery of input energy and subsequent removal of waste heat are at the heart of the data center's design and drive the vast majority of non-computing costs. These costs are proportional to the amount of power delivered and typically run in range of $10–20 per watt (see Chapter 6), but can vary considerably depending on size, location, and design.

4.1   DATA CENTER OVERVIEW

4.1.1   TIER CLASSIFICATIONS AND SPECIFICATIONS

The design of a data center is often classified using a system of four tiers [TSB]. The Uptime Institute, a professional services organization specializing in data centers, and the Telecommunications Industry Association (TIA), an industry group accredited by ANSI and comprised of approximately 400 member companies, both advocate a 4-tier classification loosely based on the power distribution, uninterruptible power supply (UPS), cooling delivery, and redundancy of the data center [UpIOS, TIA].

- *Tier I* data centers have a single path for power distribution, UPS, and cooling distribution, without redundant components.

- *Tier II* adds redundant components to this design (N + 1), improving availability.

- *Tier III* data centers have one active and one alternate distribution path for utilities. Each path has redundant components and is concurrently maintainable. Together they provide redundancy that allows planned maintenance without downtime.

- *Tier IV* data centers have two simultaneously active power and cooling distribution paths, redundant components in each path, and are supposed to tolerate any single equipment failure without impacting the load.

The Uptime Institute's specification focuses on data center performance at a high level. The specification implies topology rather than prescribing a specific list of components to meet the requirements (notable exceptions are the amount of backup diesel fuel and water storage, and ASHRAE temperature design points [UpIT]). With the Uptime standards, there are many architectures that can achieve a given tier classification. In contrast, the TIA-942 standard is more prescriptive and specifies a variety of implementation details, such as building construction, ceiling height, voltage levels, types of racks, and patch cord labeling.

Formally achieving tier classification is difficult and requires a full review from one of the certifying bodies. For this reason most data centers are not formally rated. Most commercial data centers fall somewhere between tiers III and IV, choosing a balance between construction cost and reliability. Generally, the lowest individual subsystem rating (cooling, power, and so on) determines the overall tier classification of the data center.

Real-world data center reliability is strongly influenced by the quality of the organization running the data center, not just the design. Theoretical availability estimates used in the industry range from 99.7% for tier II data centers to 99.98% and 99.995% for tiers III and IV, respectively [TIA]. However, real-world reliability often is dominated by factors not included in these calculations; for example, the Uptime Institute reports that over 70% of data center outages are the result of human error, including management decisions on staffing, maintenance, and training [UpIOS]. Furthermore, in an environment using continuous integration and delivery of software, software-induced outages dominate building outages.

4.1.2   BUILDING BASICS

Data center sizes vary widely and are commonly described in terms of either the floor area for IT equipment or *critical power*, the total power that can be continuously supplied to IT equipment; Two thirds of U.S. servers were recently housed in data centers smaller than 5,000 ft^2 (450 square meters) and with less than 1 MW of critical power [EPA07, Koo11]. Large commercial data centers are built to host servers from multiple companies (often called co-location data centers, or "colos") and can support a critical load of tens of megawatts; the data centers of large cloud providers are similar, although often larger. Many data centers are single story, while some are multi-story (Figure 4.1); the critical power of some data center buildings can exceed 100 MW today.

Figure 4.1: Google's four-story cloud data center in Mayes County, Oklahoma.

At a high level, a data center building has multiple components. There is a *mechanical yard* (or a central utility building) that hosts all the cooling systems, such as cooling towers and chillers. There is an *electrical yard* that hosts all the electrical equipment, such as generators and power distribution centers. Within the data center, the main *server hall* hosts the compute, storage, and networking equipment organized into hot aisles and cold aisles. The server floor can also host *repair areas* for operations engineers. Most data centers also have separate areas designated for *networking*, including inter-cluster, campus-level, facility management, and long-haul connectivity. Given the criticality of networking for data center availability, the networking areas typically have additional physical security and high-availability features to ensure increased reliability. The *data center building* construction follows established codes around fire-resistive and non-combustible construction, safety, and so on [IBC15], and the design also incorporates elaborate *security* for access, including circle locks, metal detectors, guard personnel, and an extensive network of cameras.

Figure 4.2 shows an aerial view of a Google data center campus in Council Bluffs, Iowa. Figure 4.3 zooms in on one building to highlight some of the typical components in greater detail.

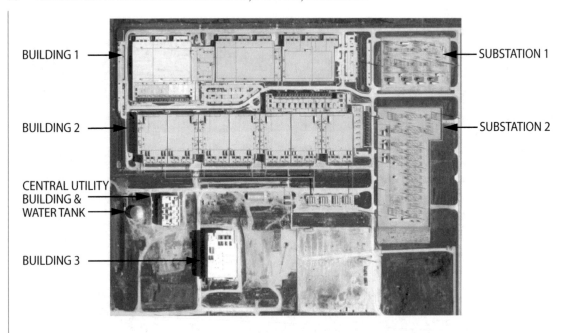

BUILDING 1

BUILDING 2

CENTRAL UTILITY
BUILDING &
WATER TANK

BUILDING 3

SUBSTATION 1

SUBSTATION 2

Figure 4.2: Aerial view of a Google data center campus in Council Bluffs, Iowa.

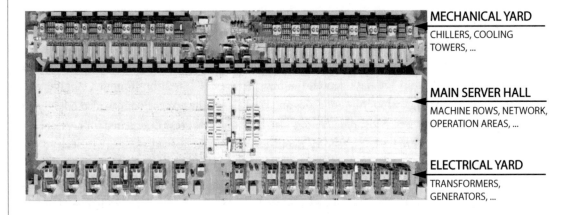

MECHANICAL YARD

CHILLERS, COOLING
TOWERS, ...

MAIN SERVER HALL

MACHINE ROWS, NETWORK,
OPERATION AREAS, ...

ELECTRICAL YARD

TRANSFORMERS,
GENERATORS, ...

Figure 4.3: A Google data center building in Council Bluffs, Iowa, showing the mechanical yard, electrical yard, and server hall.

Figure 4.4 shows the components of a typical data center architecture. Beyond the IT equipment (discussed in Chapter 3), the two major systems in the data center provide power delivery (shown in red, indicated by numbers) and cooling (shown in green, indicated by letters). We discuss these in detail next.

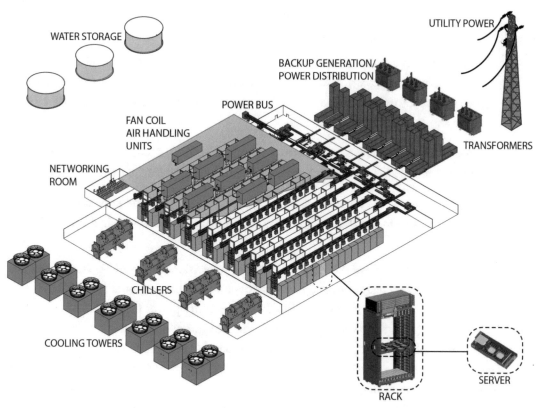

Figure 4.4: The main components of a typical data center.

4.2   DATA CENTER POWER SYSTEMS

Power enters first at a utility substation (not shown) which transforms high voltage (typically 110 kV and above) to medium voltage (typically less than 50 kV). Medium voltage is used for site-level distribution to the primary distribution centers (also known as unit substations), which include the primary switchgear and medium-to-low voltage transformers (typically below 1,000 V). From here, the power enters the building with the low-voltage lines going to the uninterruptible power supply (UPS) systems. The UPS switchgear also takes a second feed at the same voltage from a set of diesel generators that cut in when utility power fails. An alternative is to use a flywheel or alternator assembly, which is turned by an electric motor during normal operation, and couples to a diesel motor via a clutch during utility outages. In any case, the outputs of the UPS system are routed to the data center floor where they are connected to Power Distribution Units (PDUs). PDUs are the last layer in the transformation and distribution architecture and route individual circuits to the computer cabinets.

4.2.1   UNINTERRUPTIBLE POWER SYSTEMS (UPS)

The UPS typically combines three functions in one system.

- First, it contains a transfer switch that chooses the active power input (either utility power or generator power). After a power failure, the transfer switch senses when the generator has started and is ready to provide power; typically, a generator takes 10–15 s to start and assume the full rated load.

- Second, it contains some form of energy storage (electrical, chemical, or mechanical) to bridge the time between the utility failure and the availability of generator power.

- Third, it conditions the incoming power feed, removing voltage spikes or sags, or harmonic distortions in the AC feed. This conditioning can be accomplished via "double conversion."

A traditional UPS employs AC–DC–AC double conversion. Input AC is rectified to DC, which feeds a UPS-internal bus connected to strings of batteries. The output of the DC bus is then inverted back to AC to feed the data center PDUs. When utility power fails, input AC is lost but internal DC remains (from the batteries) so that AC output to the data center continues uninterrupted. Eventually, the generator starts and resupplies input AC power.

Traditional double-conversion architectures are robust but inefficient, wasting as much as 15% of the power flowing through them as heat. Newer designs such as line-interactive, delta-conversion, multi-mode, or flywheel systems operate at efficiencies in the range of 96–98% over a wide range of load cases. Additionally, "floating" battery architectures such as Google's on-board UPS [Whi+] place a battery on the output side of the server's AC/DC power supply, thus requiring only a small trickle of charge and a simple switching circuit. These systems have demonstrated efficiencies exceeding 99%. A similar strategy was later adopted by the OpenCompute UPS [OCP11], which distributes a rack of batteries for every four server racks, and by Google's high-availability rack systems, which contain servers powered from a rack-level DC bus fed from either modular, redundant rectifiers or modular, redundant battery trays.

Because UPS systems take up a sizeable amount of space, they are usually housed in a room separate from the data center floor. Typical UPS capacities range from hundreds of kilowatts up to two megawatts or more, depending on the power needs of the equipment. Larger capacities are achieved by combining several smaller units.

It's possible to use UPS systems not only in utility outages but also as supplementary energy buffers for power and energy management. We discuss these proposals further in the next chapter.

4.2.2   POWER DISTRIBUTION UNITS (PDUS)

In our example data center, the UPS output is routed to PDUs on the data center floor. PDUs resemble breaker panels in residential houses but can also incorporate transformers for final voltage adjustments. They take a larger input feed and break it into many smaller circuits that distribute power to the actual servers on the floor. Each circuit is protected by its own breaker, so a short in a server or power supply will trip only the breaker for that circuit, not the entire PDU or even the UPS. A traditional PDU handles 75–225 kW of load, whereas a traditional circuit handles a maximum of approximately 6 kW (20 or 30 A at 110–230 V). The size of PDUs found in large-scale data centers is much higher, however, corresponding to the size of the largest commodity backup generators (in megawatts), with circuits sometimes corresponding to high-power racks ranging in the tens of kW capacity. PDUs often provide additional redundancy by accepting two independent ("A" and "B") power sources and are able to switch between them with a small delay. The loss of one source does not interrupt power to the servers. In this scenario, the data center's UPS units are usually duplicated on A and B sides, so that even a UPS failure will not interrupt server power.

In North America, the input to the PDU is commonly 480 V 3-phase power. This requires the PDU to perform a final transformation step to deliver the desired 110 V output for the servers, thus introducing another source of inefficiency. In the EU, input to the PDU is typically 400 V 3-phase power. By taking power from any single phase to neutral combination, it is possible to deliver a desirable 230 V without an extra transformer step. Using the same trick in North America requires computer equipment to accept 277 V (as derived from the 480 V input to the PDU), which unfortunately exceeds the upper range of standard power supplies.

Real-world data centers contain many variants of the simplified design described here. These include the "paralleling" of generators or UPS units, an arrangement where multiple devices feed a shared bus so the load of a failed device can be picked up by other devices, similar to handling disk failures in a RAID system. Common paralleling configurations include N + 1 (allowing one failure or maintenance operation at a time), N + 2 (allowing one failure even when one unit is offline for maintenance), and 2N (fully redundant pairs).

4.2.3   COMPARISON OF AC AND DC DISTRIBUTION ARCHITECTURES

The use of high-voltage DC (HVDC) on the utility grid presents advantages for connecting incompatible power grids, providing resistance to cascading failures, and long-distance transmission efficiency. In data centers, the case for DC distribution is centered around efficiency improvements, increased reliability from reduced component counts, and easier integration of distributed generators with native DC outputs. In comparison with the double-conversion UPS mentioned above, DC systems eliminate the final inversion step of the UPS. If the voltage is selected to match the DC primary stage of the server power supply unit (PSU), three additional steps are eliminated: PDU transformation, PSU rectification, and PSU power factor correction.

Figure 4.5 compares AC and DC distribution architectures commonly used in the data center industry. State-of-the-art, commercially available efficiencies (based on [GF]) are shown for each stage of the "power train." The overall power train efficiency using state-of-the-art components remains a few percent higher for DC distribution as compared to AC distribution; this difference was more pronounced in data centers with older components [Pra+06]. Note that the AC architecture shown corresponds to the voltage scheme commonly found in North America; in most other parts of the world the additional voltage transformation in the AC PDU can be avoided, leading to slightly higher PDU efficiency.

CONVENTIONAL AC ARCHITECTURE

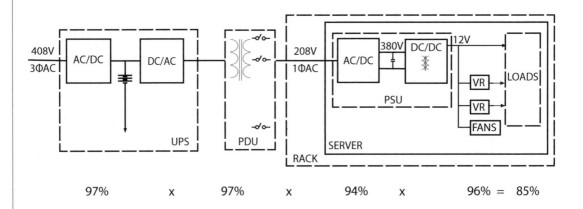

CONVENTIONAL DC ARCHITECTURE

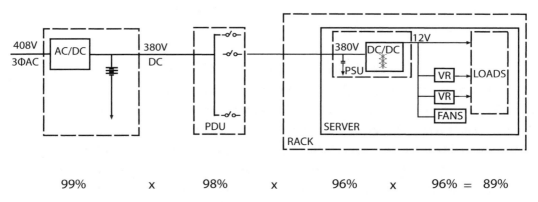

Figure 4.5: Comparison of AC and DC distribution architectures commonly employed in the data center industry.

The placement of batteries as parallel sources near the load can virtually eliminate UPS losses: ~0.1% compared to ~1%–3% for an in-line UPS with 480 V AC input. In typical Google designs, batteries are included either on server trays or as modules of a rack in parallel with a DC bus, and this has allowed elimination of the upstream UPS, further increasing the power train efficiency.

Commercial DC equipment for data centers is available, but costs remain higher than for comparable AC equipment. Similarly, the construction of large data centers involves hundreds and sometimes thousands of skilled workers. While only a subset of these will be electricians, the limited availability of DC technicians may lead to increased construction, service, and operational costs. However, DC power distribution is more attractive when integrating distributed power generators such as solar photovoltaic, fuel cells, and wind turbines. These power sources typically produce native DC and are easily integrated into a DC power distribution architecture.

4.3   EXAMPLE: RADIAL POWER DISTRIBUTION WITH REDUNDANCY

A conventional AC power distribution scheme for a large data center is shown in Figure 4.6. This topology is known as "radial" because power fans out to the entire data center floor from a pair of medium voltage buses that provide redundancy in case of loss of a utility feed. Low voltage (400–480 V) AC power is supplied to the data center floor by many PDUs, each fed by either a step-down transformer for utility power or a backup generator. In addition, power availability is greatly enhanced by an isolated redundant PDU. This module is identical to the others, but needs to carry load only when other low voltage equipment fails or needs to be taken temporarily out of service.

4.4   EXAMPLE: MEDIUM VOLTAGE POWER PLANE

An interesting modern architecture for data center power distribution is Google's medium voltage power plane (Figure 4.7), which allows for sharing of power across the data center. High availability at the building level is provided by redundant utility AC inputs. Building-level transformers step the voltage down to a medium voltage of 11–15 kV for further distribution through the building's electrical rooms. For backup power, a "farm" of many medium voltage generators are paralleled to a bus, and automated systems consisting of breakers and switches select between the utility and generator sources. Redundant paths exist from both utility and generator sources to many unit sub-stations. Each unit substation steps down the voltage to approximately 400 V AC for distribution to a row of racks on the data center floor.

The power plane architecture offers several advantages with respect to traditional radial architectures. First, a large pool of diverse workloads can increase the opportunity for power over-subscription [Ran+06], discussed in the following chapter. Roughly speaking, Google's power plane architecture doubles the quantity of IT equipment that can be deployed above and beyond a data

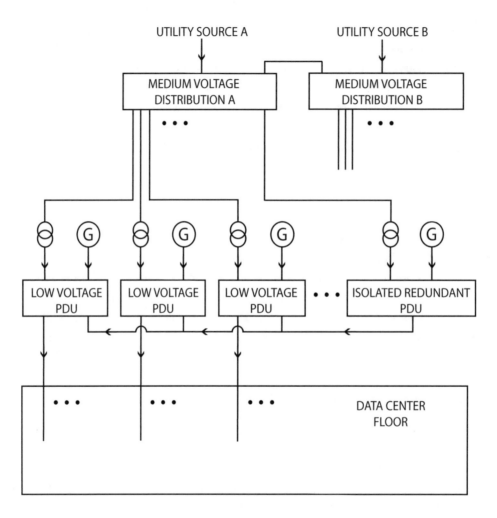

Figure 4.6: A radial power architecture with generators distributed among many low voltage PDUs. Each PDU has its own backup generator indicated by a "G." Power loss due to failures of low voltage equipment is greatly mitigated by the presence of an isolated redundant PDU, which can take the place of any other PDU as a low voltage source.

center's critical power capacity. Second, the generator farm offers resilience against generator failures with a minimum of redundant equipment. Finally, power is more fungible across the entire data center floor: with appropriate sizing of both medium- and low-voltage distribution components, a high dynamic range of deployment power density can be supported without stranding power. This is an important benefit given that rack power varies substantially depending on the type of IT equipment within the rack. For example, storage-intensive racks consume much less power than compute-intensive racks. With traditional radial power architectures, a low power density in one region of the data center floor can result in permanently underused infrastructure. The medium-voltage power plane enables power sharing across the floor: high power racks in one region can compensate for low power racks in another region, ensuring full utilization of the building's power capacity.

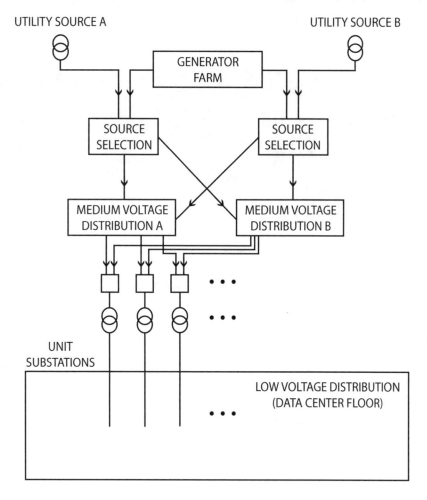

Figure 4.7: Concept for medium-voltage power plane architecture.

4.5   DATA CENTER COOLING SYSTEMS

Data center cooling systems remove the heat generated by the equipment. To remove heat, a cooling system must employ some hierarchy of loops, each circulating a cold medium that warms up via some form of heat exchange and is somehow cooled again. An open loop replaces the outgoing warm medium with a cool supply from the outside, so that each cycle through the loop uses new material. A closed loop recirculates a separate medium, continuously transferring heat to either another loop via a heat exchanger or to the environment; all systems of loops must eventually transfer heat to the outside environment.

The simplest topology is fresh air cooling (or air economization)—essentially, opening the windows. Such a system is shown in Figure 4.8. This is a single, open-loop system that we discuss in more detail in the section on free cooling.

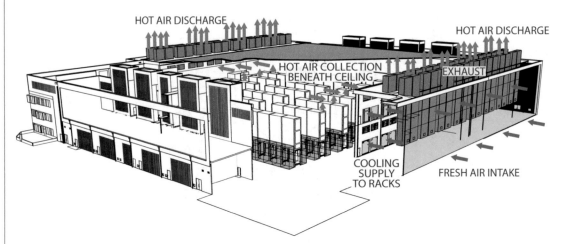

Figure 4.8: Airflow schematic of an air-economized data center.

Closed-loop systems come in many forms, the most common being the air circuit on the data center floor. Its function is to isolate and remove heat from the servers and transport it to a heat exchanger. As shown in Figure 4.9, cold air flows to the servers, heats up, and eventually reaches a heat exchanger to cool it down again for the next cycle through the servers.

Typically, data centers employ raised floors, concrete tiles installed onto a steel grid resting on stanchions two to four feet above the slab floor. The underfloor area often contains power cables to racks, but its primary purpose is to distribute cool air to the server racks. The airflow through the underfloor plenum, the racks, and back to the CRAC (a 1960s term for *computer room air conditioning*) defines the primary air circuit.

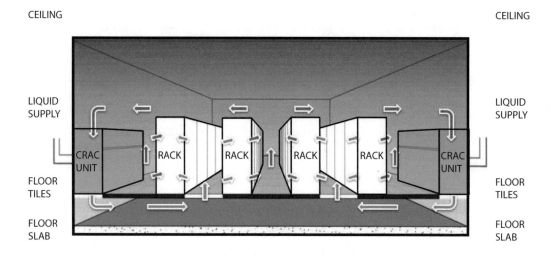

Figure 4.9: Raised floor data center with hot-cold aisle setup (image courtesy of DLB Associates [Dye06]).

The simplest closed-loop systems contain two loops. The first loop is the air circuit shown in Figure 4.9, and the second loop (the liquid supply inside the CRACs) leads directly from the CRAC to external heat exchangers (typically placed on the building roof) that discharge the heat to the environment.

A three-loop system commonly used in large-scale data centers is shown in Figure 4.10. The first *datacenter floor loop* involves circulating air that is alternately cooled by fan coils and heated by IT equipment on the data center floor. In the *process loop*, warm water from the fan coils returns to the cooling plant to be chilled and pumped back to the fan coils. Finally, the *condenser water loop* removes heat received from the process water through a combination of mechanical refrigeration by chiller units and evaporation in cooling towers; the condenser loop is so named because it removes heat from the condenser side of the chiller. Heat exchangers perform much of the heat transfer between the loops, while preventing process water from mixing with condenser water.

Each topology presents tradeoffs in complexity, efficiency, and cost. For example, fresh air cooling can be very efficient but does not work in all climates, requires filtering of airborne particulates, and can introduce complex control problems. Two-loop systems are easy to implement, relatively inexpensive to construct, and offer isolation from external contamination, but typically have lower operational efficiency. A three-loop system is the most expensive to construct and has moderately complex controls, but offers contaminant protection and good efficiency when employing economizers.

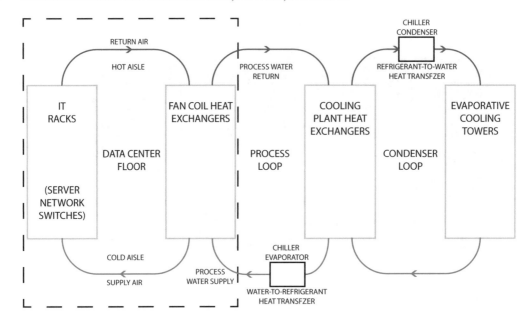

Figure 4.10: Three-loop data center cooling system. (Note that in favorable weather conditions, the entire data center heat load can be removed by evaporative cooling of the condenser water; the chiller evaporator and chiller condenser heat transfer steps then become unnecessary.)

Additionally, generators (and sometimes UPS units) provide backup power for most mechanical cooling equipment because the data center may overheat in a matter of minutes without cooling. In a typical data center, chillers and pumps can add 40% or more to the critical load supported by generators, significantly adding to the overall construction cost.

CRACs, chillers, and cooling towers are among the most important building blocks in data center cooling systems, and we take a slightly closer look at each below.

4.5.1   COMPUTER ROOM AIR CONDITIONERS (CRACS)

All CRACs contain a heat exchanger, air mover, and controls. They mostly differ by the type of cooling they employ:

- direct expansion (DX);

- fluid solution; and

- water.

A DX unit is a split air conditioner with cooling (evaporator) coils inside the CRAC, and heat-rejecting (condenser) coils outside the data center. The fluid solution CRAC shares this basic architecture but circulates a mixture of water and glycol through its coils rather than a phase-change refrigerant. Finally, a water-cooled CRAC connects to a chilled water loop.

CRAC units pressurize the raised floor plenum by blowing cold air into the underfloor space, which then escapes through perforated tiles in front of the server racks. The air flows through the servers and is expelled into a "hot aisle." Racks are typically arranged in long rows that alternate between cold and hot aisles to reduce inefficiencies caused by mixing hot and cold air. In fact, many newer data centers physically isolate the cold or hot aisles with walls [PF]. As shown in Figure 4.9, the hot air produced by the servers recirculates back to the intakes of the CRACs, where it is cooled and exhausted into the raised floor plenum again.

4.5.2   CHILLERS

A water-cooled chiller as shown in Figure 4.11 can be thought of as a water-cooled air conditioner.

Figure 4.11: Water-cooled centrifugal chiller.

Chillers submerge the evaporator and condenser coils in water in two large, separate compartments joined via a top-mounted refrigeration system consisting of a compressor, expansion valve, and piping. In the cold compartment, warm water from the data center is cooled by the evaporator coil prior to returning to the process chilled water supply (PCWS) loop. In the hot compartment, cool water from the condenser water loop is warmed by the condenser coil and carries the heat away to the cooling towers where it is rejected to the environment by evaporative cooling. Because the chiller uses a compressor, a significant amount of energy is consumed to perform its work.

4.5.3   COOLING TOWERS

Cooling towers (Figure 4.12) cool a water stream by evaporating a portion of it into the atmosphere. The energy required to change the liquid into a gas is known as the latent heat of vaporization, and the temperature of the water can be dropped significantly given favorable dry conditions. The water flowing through the tower comes directly from the chillers or from another heat exchanger connected to the PCWS loop. Figure 4.13 illustrates how it works.

Figure 4.12: Data center cooling towers.

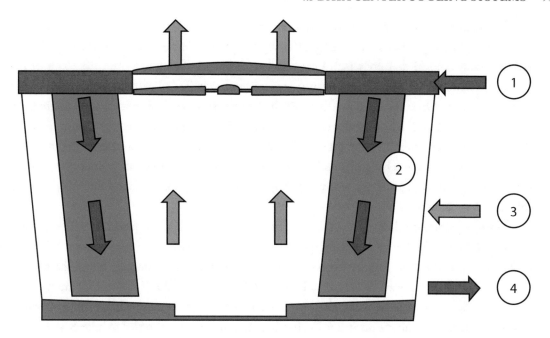

Figure 4.13: How a cooling tower works. The numbers correspond to the associated discussions in the text.

1. Hot water from the data center flows from the top of the cooling tower onto "fill" material inside the tower. The fill creates additional surface area to improve evaporation performance.

2. As the water flows down the tower, some of it evaporates, drawing energy out of the remaining water and reducing its temperature.

3. A fan on top draws air through the tower to aid evaporation. Dry air enters the sides and humid air exits the top.

4. The cool water is collected at the base of the tower and returned to the data center.

Cooling towers work best in temperate climates with low humidity; ironically, they do not work as well in very cold climates because they need additional mechanisms to prevent ice formation on the towers and in the pipes.

4.5.4   FREE COOLING

Free cooling refers to the use of cold outside air to either help produce chilled water or directly cool servers. It is not completely free in the sense of zero cost, but it involves very low-energy costs compared to chillers.

As mentioned above, air-economized data centers are open to the external environment and use low dry bulb temperatures for cooling. (The dry bulb temperature is the air temperature measured by a conventional thermometer). Large fans push outside air directly into the room or the raised floor plenum when outside temperatures are within limits (for an extreme experiment in this area, see [AM08]). Once the air flows through the servers, it is expelled outside the building. An air-economized system can be very efficient but requires effective filtering to control contamination, may require auxiliary cooling (when external conditions are not favorable), and may be difficult to control. Specifically, if there is a malfunction, temperatures will rise very quickly since air can store relatively little heat. By contrast, a water-based system can use a water storage tank to provide a significant thermal buffer.

Water-economized data centers take advantage of the wet bulb temperature [Wbt]. The wet bulb temperature is the lowest water temperature that can be reached by evaporation. The dryer the air, the bigger the difference between dry bulb and wet bulb temperatures; the difference can exceed 10°C, and thus a water-economized data center can run without chillers for many more hours per year. For this reason, some air-economized data centers employ a hybrid system where water is misted into the airstream (prior to entering the data center) in order to take advantage of evaporation cooling.

Typical water-economized data centers employ a parallel heat exchanger so that the chiller can be turned off when the wet bulb temperature is favorable. Depending on the capacity of the cooling tower (which increases as the wet bulb temperature decreases), a control system balances water flow between the chiller and the cooling tower.

Yet another approach uses a radiator instead of a cooling tower, pumping the condenser fluid or process water through a fan-cooled radiator. Similar to the glycol/water-based CRAC, such systems use a glycol-based loop to avoid freezing. Radiators work well in cold climates (say, a winter in Chicago) but less well at moderate or warm temperatures because the achievable cold temperature is limited by the external dry bulb temperature, and because convection is less efficient than evaporation.

4.5.5   AIR FLOW CONSIDERATIONS

Most data centers use the raised floor setup discussed above. To change the amount of cooling delivered to a particular rack or row, we exchange perforated tiles with solid tiles or vice versa. For cooling to work well, the cold airflow coming through the tiles should match the horizontal airflow through the servers in the rack. For example, if a rack has 10 servers with an airflow of 100 cubic feet per minute (CFM) each, then the net flow out of the perforated tile should be 1,000 CFM (or higher if the air path to the servers is not tightly controlled). If it is lower, some of the servers will receive cold air while others will ingest recirculated warm air from above the rack or other leakage paths.

Figure 4.14 shows the results of a Computational Fluid Dynamics (CFD) analysis for a rack that is oversubscribing the data center's airflow.

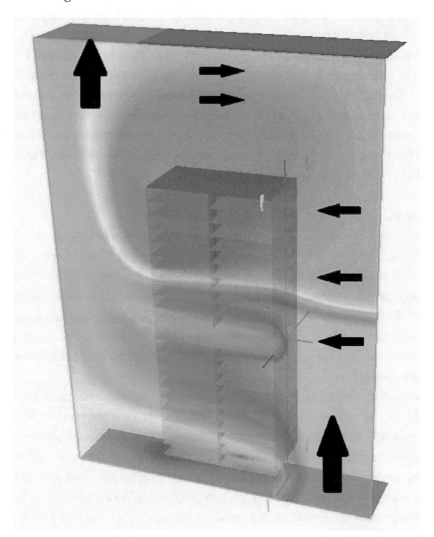

Figure 4.14: CFD model showing recirculation paths and temperature stratification for a rack with under-provisioned airflow.

In this example, recirculation across the top of the rack causes the upper servers to ingest warm air. The servers on the bottom are also affected by a recirculation path under the rack. Blockages from cable management hardware cause a moderate warm zone about halfway up the rack.

The facility manager's typical response to such a situation is to lower the temperature of the CRAC output. That works, but increases energy costs significantly, so it's better to fix the underlying problem instead and physically separate cold and warm air as much as possible, while optimizing the path back to the CRACs. In this setup the entire room is filled with cool air (because the warm exhaust is kept inside a separate plenum or duct system) and, thus, all servers in a rack will ingest air at the same temperature [PF].

Air flow limits the power density of data centers. For a fixed temperature differential across a server, a rack's airflow requirement increases with power consumption, and the airflow supplied via the raised floor tiles must increase linearly with power. That in turn increases the amount of static pressure needed in the underfloor plenum. At low power densities, this is easy to accomplish, but at some point the laws of physics start to make it economically impractical to further increase pressure and airflow. Typically, these limitations make it hard to exceed power densities of more than 150–200 W/sq ft without substantially increased cost.

4.5.6   IN-RACK, IN-ROW, AND LIQUID COOLING

In-rack cooling can increase power density and cooling efficiency beyond the conventional raised-floor limit. Typically, an in-rack cooler adds an air-to-water heat exchanger at the back of a rack so the hot air exiting the servers immediately flows over coils cooled by water, essentially short-circuiting the path between server exhaust and CRAC input. In-rack cooling might remove part or all of the heat, effectively replacing the CRACs. Obviously, chilled water needs to be brought to each rack, greatly increasing the cost of plumbing. Some operators may also worry about having water on the data center floor, since leaky coils or accidents might cause water to spill on the equipment.

In-row cooling works like in-rack cooling except the cooling coils aren't in the rack, but adjacent to the rack. A capture plenum directs the hot air to the coils and prevents leakage into the cold aisle. Figure 4.15 shows an in-row cooling product and how it is placed between racks.

Finally, we can directly cool server components using cold plates, that is, local, liquid-cooled heat sinks. It is usually impractical to cool all compute components with cold plates. Instead, components with the highest power dissipation (such as processor chips) are targeted for liquid cooling while other components are air-cooled. The liquid circulating through the heat sinks transports the heat to a liquid-to-air or liquid-to-liquid heat exchanger that can be placed close to the tray or rack, or be part of the data center building (such as a cooling tower).

Figure 4.15: In-row air conditioner.

In spite of the higher cost and mechanical design complexity, cold plates are becoming essential for cooling very high-density workloads whose TDP per chip exceeds what is practical to cool with regular heatsinks (typically, 200–250 W per chip). A recent example is Google's third-generation tensor processing unit (TPU): as shown in Figure 4.16, four TPUs on the same motherboard are cooled in series on a single water loop.

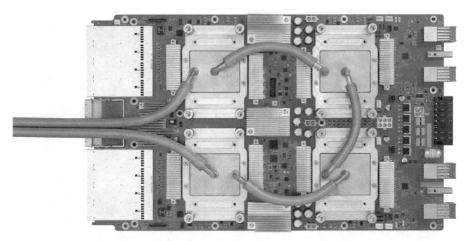

Figure 4.16: Copper cold plates and hose connections provide liquid cooling for Google's third-generation TPU.

4.5.7   CONTAINER-BASED DATA CENTERS

Container-based data centers go one step beyond in-row cooling by placing the server racks inside a container (typically 20 or 40 ft long) and integrating heat exchange and power distribution into the container as well. Similar to in-row cooling, the container needs a supply of chilled water and uses coils to remove all heat. Close-coupled air handling typically allows higher power densities than regular raised-floor data centers. Thus, container-based data centers provide all the functions of a typical data center room (racks, CRACs, PDU, cabling, lighting) in a small package. Figure 4.17 shows an isometric cutaway of Google's container design.

Like a regular data center room, containers must be accompanied by outside infrastructure such as chillers, generators, and UPS units to be fully functional.

To our knowledge, the first container-based data center was built by Google in 2005 [GInc09], and the idea dates back to a Google patent application in 2003. However, subsequent generations of Google data centers have moved away from containers and instead incorporate the same principles at a broader warehouse level. Some other large-scale operators, including Microsoft [Micro] and eBay [eBay12], have also reported using containers in their facilities, but today they are uncommon.

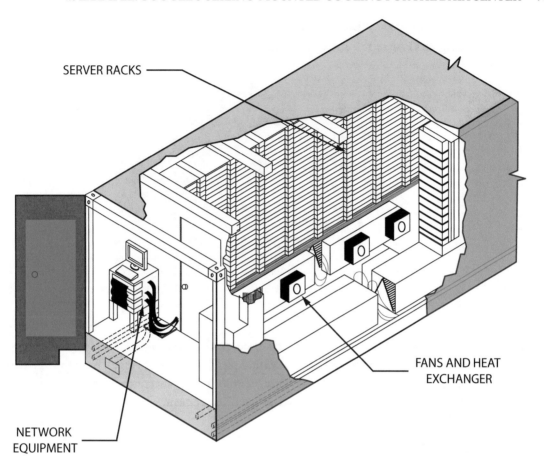

SERVER RACKS

FANS AND HEAT
EXCHANGER

NETWORK
EQUIPMENT

Figure 4.17: Google's container design includes all the infrastructure of the data center floor.

4.6   EXAMPLE: GOOGLE'S CEILING-MOUNTED COOLING FOR THE DATA CENTER

Figure 4.18 illustrates the main features and air flow of Google's overhead cooling system. This represents one variation on the efficient hot aisle containment that has become prevalent in the data center industry. Tall, vertical hot aisle plenums duct the exhaust air from the rear of the IT racks to overhead fan coils. The fan coils receive chilled process water from an external cooling plant; this water flows through multiple tube passages attached to fins, absorbing heat from the incoming hot air. Blowers in the fan coil units force the cooled air downward into the cold aisle, where it enters the intakes of servers and networking equipment. Together with the cooling plant and process water loops, this air loop comprises a highly-efficient, end-to-end cooling system that consumes energy amounting to <10% of the energy consumed by the IT equipment.

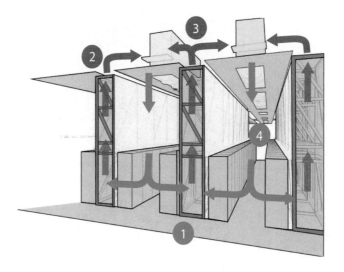

Figure 4.18: Cross-sectional view of a cold aisle and associated hot air plenums in a Google data center. (1) Hot exhaust from IT equipment rises in a vertical plenum space. (2) Hot air enters a large plenum space above the drop ceiling. (3) Heat is exchanged with process water in a fan coil unit, which also (4) blows the cold air down toward the intake of the IT equipment.

4.7   SUMMARY

Data centers power the servers they contain and remove the heat generated. Historically, data centers have consumed twice as much energy as needed to power the servers, but when best practices are employed this overhead shrinks to 10–20%. Key energy saving techniques include free-cooling (further boosted by raising the target inlet temperature of servers), well-managed air flow, and high-efficiency power distribution and UPS components.

CHAPTER 5

Energy and Power Efficiency

Energy efficiency has been a major technology driver in the mobile and embedded areas for a long time. Work in this area originally emphasized extending battery life, but then expanded to include reducing peak power because thermal constraints began to limit further CPU performance improvements or packaging density in small devices. However, energy management is also a key issue for servers and data center operations, one that focuses on reducing all energy-related costs, including capital and operating expenses as well as environmental impacts. Many energy-saving techniques developed for mobile devices are natural candidates for tackling this new problem space, but ultimately a WSC is quite different from a mobile device. In this chapter, we describe some of the most relevant aspects of energy and power efficiency for WSCs, starting at the data center level and continuing to component-level issues.

5.1 DATA CENTER ENERGY EFFICIENCY

The broadest definition of WSC energy efficiency would measure the energy used to run a particular workload (say, to sort a petabyte of data). Unfortunately, no two companies run the same workloads and, as discussed in Chapter 2, real-world application mixes change all the time, so it is hard to benchmark WSCs this way. Thus, even though such benchmarks have been contemplated [Riv+07], they haven't yet been widely used [TGGb]. However, it is useful to view energy efficiency as the product of three factors we can independently measure and optimize:

$$\text{Efficiency} = \frac{\text{Computation}}{\text{Total Energy}} = \underbrace{\left(\frac{1}{\text{PUE}}\right)}_{(a)} \times \underbrace{\left(\frac{1}{\text{SPUE}}\right)}_{(b)} \times \underbrace{\left(\frac{\text{Computation}}{\text{Total Energy to Electronic Components}}\right)}_{(c)}.$$

In this equation, the first term (a) measures facility efficiency, the second (b) measures server power conversion efficiency, and the third (c) measures the server's architectural efficiency. We discuss these factors in the following sections.

5.1.1 THE PUE METRIC
Power usage effectiveness (PUE) reflects the quality of the data center building infrastructure itself [TGGc], and captures the ratio of total building power to IT power (the power consumed by the computing, networking, and other IT equipment). IT power is sometimes referred to as "critical power."

$$PUE = (Facility\ power)\ /\ (IT\ Equipment\ power).$$

PUE has gained a lot of traction as a data center efficiency metric since widespread reporting started on it around 2009. We can easily measure PUE by adding electrical meters to the lines powering the various parts of a data center, thus determining how much power is used by chillers and UPSs.

Historically, the PUE for the average data center has been embarrassingly poor. According to a 2006 study [MB06], 85% of data centers were estimated to have a PUE greater than 3.0. In other words, the building's mechanical and electrical systems consumed twice as much power as the actual computing load. Only 5% had a PUE of 2.0 or better.

A subsequent EPA survey of over 100 data centers reported an average PUE of 1.91 [PUE10]. A few years back, an Uptime Institute survey of over 1,100 data centers covering a range of geographies and sizes reported an average PUE value between 1.8 and 1.89 [UpI12, Hes14]. More recently, a 2016 report from LBNL noted PUEs of 1.13 for hyperscale data centers (warehouse-scale computers) and 1.6–2.35 for traditional data centers [She+16]. Figure 5.1 shows the distribution of results from one of these studies [UpI12]. Cold and hot aisle containment and increased cold aisle temperature are the most common improvements implemented. Large facilities reported the biggest improvements, and about half of small data centers (with less than 500 servers) still were not measuring PUE.

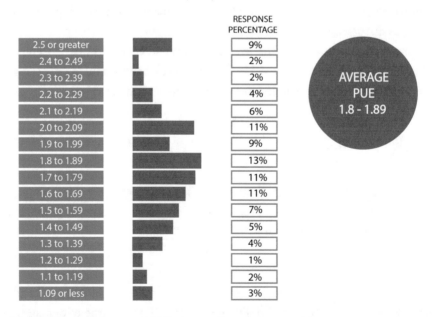

AVERAGE PUE OF LARGEST DATA CENTER

	RESPONSE PERCENTAGE
2.5 or greater	9%
2.4 to 2.49	2%
2.3 to 2.39	2%
2.2 to 2.29	4%
2.1 to 2.19	6%
2.0 to 2.09	11%
1.9 to 1.99	9%
1.8 to 1.89	13%
1.7 to 1.79	11%
1.6 to 1.69	11%
1.5 to 1.59	7%
1.4 to 1.49	5%
1.3 to 1.39	4%
1.2 to 1.29	1%
1.1 to 1.19	2%
1.09 or less	3%

AVERAGE PUE 1.8 - 1.89

Figure 5.1: Uptime Institute survey of PUE for 1100+ data centers. This detailed data is based on a 2012 study [UpI12] but the trends are qualitatively similar to more recent studies (e.g., 2016 LBNL study [She+16]).

Very large operators (usually consumer internet companies like Google, Microsoft, Yahoo!, Facebook, Amazon, Alibaba, and eBay) have reported excellent PUE results over the past few years, typically below 1.2, although only Google has provided regular updates of its entire fleet based on a clearly defined metric (Figure 5.2) [GDCa]. At scale, it is easy to justify the importance of efficiency; for example, Google reported having saved over one billion dollars to date from energy efficiency measures [GGr].

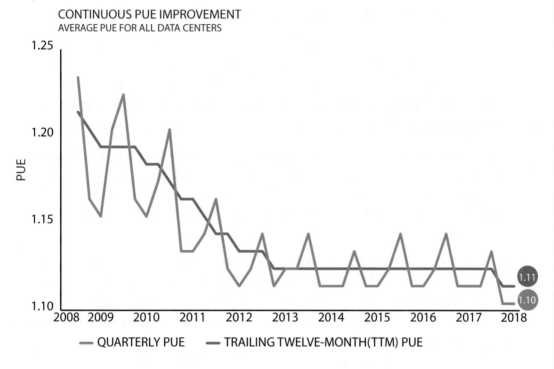

Figure 5.2: PUE data for all large-scale Google data centers over time [GDCa].

5.1.2   ISSUES WITH THE PUE METRIC

Although The Green Grid (TGG) publishes detailed guidelines on how to measure and report PUE [TGGd], many published values aren't directly comparable, and sometimes PUE values are used in marketing documents to show best-case values that aren't real. The biggest factors that can skew PUE values are as follows.

- Not all PUE measurements include the same overheads. For example, some may include losses in the primary substation transformers or in wires feeding racks from PDUs, whereas others may not. Google reported a fleet-wide PUE of 1.12 using a comprehensive definition of overhead that includes all known sources, but could have reported a PUE of 1.06 with a more "optimistic" definition of overhead [GDCb]. For PUE to be a useful metric, data center owners and operators should adhere to Green Grid guidelines [TGGd] in measurements and reporting, and be transparent about the methods used in arriving at their results.

- Instantaneous PUEs differ from average PUEs. Over the course of a day or a year, a facility's PUE can vary considerably. For example, on a cold day it might be low, but during the summer it might be considerably higher. Generally speaking, annual averages are more useful for comparisons.

- Some PUEs aren't real-world measurements. Often vendors publish "design" PUEs that are computed using optimal operating conditions and nominal performance values, or they publish a value measured during a short load test under optimal conditions. Typically, PUE values provided without details fall into this category.

- Some PUE values have higher error bars because they're based on infrequent manual readings, or on coarsely placed meters that force some PUE terms to be estimated instead of measured. For example, if the facility has a single meter measuring the critical load downstream of the UPS, PDU, and low-voltage distribution losses will need to be estimated.

In practice, PUE values should be measured in real time. Not only does this provide a better picture of diurnal and seasonal variations, it also allows the operator to react to unusual readings during day-to-day operations. For example, someone may have left on a set of backup pumps after a periodic test. With real-time metrics the operations team can quickly correct such problems after comparing expected vs. actual PUE values.

The PUE metric has been criticized as not always indicating better energy performance, because PUEs typically worsen with decreasing load. For example, assume a data center's PUE is 2.0 at a 500 kW load vs. 1.5 at a 1 MW load. If it's possible to run the given workload with a 500 kW load (for example, with newer servers), that clearly is more energy efficient despite the inferior PUE. However, this criticism merely points out that PUE is just one of three factors in the efficiency equation shown earlier in this chapter, and overall the widespread adoption of PUE measurements has arguably been the driver of the biggest improvements in data center efficiency in the past 50 years.

5.1.3 SOURCES OF EFFICIENCY LOSSES IN DATA CENTERS

The section on data center power systems in Chapter 4 describes the efficient transformation of power as it approaches the data center floor. The first two transformation steps bring the incoming high-voltage power (110 kV and above) to medium-voltage distribution levels (typically less than

50 kV) and, closer to the server floor to low voltage (typically 480 V in North America). Both steps should be very efficient, with losses typically below half a percent for each step. Inside the building, conventional double-conversion UPSs cause the most electrical loss. In the first edition we listed efficiencies of 88–94% under optimal load, significantly less if partially loaded (which is the common case). Rotary UPSs (flywheels) and high-efficiency UPSs can reach efficiencies of about 97%. The final transformation step in the PDUs accounts for an additional half-percent loss. Finally, 1–3% of power can be lost in the cables feeding low-voltage power (110 or 220 V) to the racks (recall that a large facility can have a raised floor area that is over 100 m long or wide, so power cables can be quite long).

A significant portion of data center inefficiencies stems from cooling overhead, with chillers being the largest culprit. Cooling losses are three times greater than power losses, presenting the most promising target for efficiency improvements: if all cooling losses were eliminated, the PUE would drop to 1.26, whereas a zero-loss UPS system would yield a PUE of only 1.8. Typically, the worse a facility's PUE is, the higher the percentage of the total loss comes from the cooling system [BM06]. Intuitively, there are only so many ways to mishandle a power distribution system, but many more ways to mishandle cooling.

Conversely, there are many non-intuitive ways to improve the operation of the data center's cooling infrastructure. The energy for running the cooling infrastructure has a nonlinear relationship with many system parameters and environmental factors, such as the total system load, the total number of chillers operating, and the outside wind speed. Most people find it difficult to intuit the relationship between these variables and total cooling power. At the same time, a large amount of data is being collected regularly from a network of sensors used to operate the control loop for data center cooling. The existence of this large data set suggests that machine learning and artificial intelligence could be used to find additional PUE efficiencies [EG16].

Figure 5.3 shows the typical distribution of energy losses in a WSC data center. Much of this inefficiency is caused by a historical lack of attention to power loss, not by inherent limitations imposed by physics. Less than 10 years ago, PUEs weren't formally used and a total overhead of 20% was considered unthinkably low, yet as of 2018 Google reported a fleet-wide annual average overhead of 11% [GDCb] and many others are claiming similar values for their newest facilities. However, such excellent efficiency is still confined to a small set of data centers, and many small data centers probably haven't improved much.

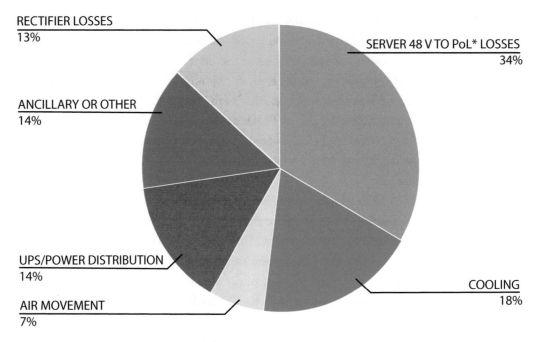

Figure 5.3: A representative end-to-end breakdown of energy losses in a typical datacenter. Note that this breakdown does not include losses of up to a few percent due to server fans or electrical resistance on server boards.

5.1.4 IMPROVING THE ENERGY EFFICIENCY OF DATA CENTERS

As discussed in the previous chapter, careful design for efficiency can substantially improve PUE [Nel+, PGE, GMT06]. To summarize, the key steps are as follows.

- *Careful air flow handling*: Isolate hot air exhausted by servers from cold air, and keep the path to the cooling coil short so that little energy is spent moving cold or hot air long distances.

- *Elevated temperatures*: Keep the cold aisle at 25–30°C rather than 18–20°C. Higher temperatures make it much easier to cool data centers efficiently. Virtually no server or network equipment actually needs intake temperatures of 20°C, and there is no evidence that higher temperatures cause more component failures [PWB07, SPW09, ES+].

- *Free cooling*: In most moderate climates, free cooling can eliminate the majority of chiller runtime or eliminate chillers altogether.

- *Better power system architecture*: UPS and power distribution losses can often be greatly reduced by selecting higher-efficiency gear, as discussed in the previous chapter.

- *Machine learning*: Apply novel machine learning techniques to discover non-intuitive techniques for controlling data center infrastructure to further reduce cooling requirements. Large amounts of data are being collected by many sensors in the data center, making this problem a natural fit for machine learning.

In April 2009, Google first published details on its data center architecture, including a video tour of a container-based data center built in 2005 [GInc09]. In 2008, this data center achieved a state-of-the-art annual PUE of 1.24, yet differed from conventional data centers only in the application of the principles listed above. Today, large-scale data centers commonly feature PUEs below 1.2, especially those belonging to cloud operators. Even in unfavorable climates, today's PUEs are lower than the state-of-the-art PUEs in 2008 For example, Google's data center in Singapore, where the average monthly temperature rarely falls below 25°C, the annual PUE is 1.18.

5.1.5 BEYOND THE FACILITY

Recall the energy efficiency formula from the beginning of this chapter:

$$\text{Efficiency} = \frac{\text{Computation}}{\text{Total Energy}} = \underbrace{\left(\frac{1}{\text{PUE}}\right)}_{(a)} \times \underbrace{\left(\frac{1}{\text{SPUE}}\right)}_{(b)} \times \underbrace{\left(\frac{\text{Computation}}{\text{Total Energy to Electronic Components}}\right)}_{(c)}.$$

So far we've discussed the first term, facility overhead. The second term (b) accounts for overheads inside servers or other IT equipment using a metric analogous to PUE: server PUE (SPUE). SPUE consists of the ratio of total server input power to its useful power, where useful power includes only the power consumed by the electronic components directly involved in the computation: motherboard, disks, CPUs, DRAM, I/O cards, and so on. Substantial amounts of power may be lost in the server's power supply, voltage regulator modules (VRMs), and cooling fans. As discussed in Chapter 4, the losses inside the server can exceed those of the entire upstream data center power train.

SPUE measurements aren't standardized like PUE but are fairly straightforward to define. Almost all equipment contains two transformation steps: the first step transforms input voltage (typically 110–220 VAC) to local DC current (typically 12 V), and in the second step VRMs transform that down to much lower voltages used by a CPU or DRAM. (The first step requires an additional internal conversion within the power supply, typically to 380 VDC.) SPUE ratios of 1.6–1.8 were common a decade ago; many server power supplies were less than 80% efficient, and many motherboards used VRMs that were similarly inefficient, losing more than 25% of input power in electrical conversion losses. In contrast, commercially available AC-input power supplies today achieve 94% efficiency, and VRMs achieve 96% efficiency (see Chapter 4). Thus, a state-of-the-art SPUE is 1.11 or less [Cli]. For example, instead of the typical 12 VDC voltage, Google uses 48 VDC voltage rack distribution system, which reduces energy losses by over 30%.

The product of PUE and SPUE constitutes an accurate assessment of the end-to-end electromechanical efficiency of a WSC. A decade ago the true (or total) PUE metric (TPUE), defined as PUE * SPUE, stood at more than 3.2 for the average data center; that is, for every productive watt, at least another 2.2 W were consumed. By contrast, a modern facility with an average PUE of 1.11 as well as an average SPUE of 1.11 achieves a TPUE of 1.23. Close attention to cooling and power system design in combination with new technology has provided an order of magnitude reduction in overhead power consumption.

5.2 THE ENERGY EFFICIENCY OF COMPUTING

So far we have discussed efficiency in electromechanical terms, the (a) and (b) terms of the efficiency equation, and largely ignored term (c), which accounts for how the electricity delivered to electronic components is actually translated into useful work. In a state-of-the-art facility, the electromechanical components have a limited potential for improvement: Google's TPUE of approximately 1.23 means that even if we eliminated all electromechanical overheads, the total energy efficiency would improve by only 19%. In contrast, the energy efficiency of computing has doubled approximately every 1.5 years in the last half century [Koo+11]. Although such rates have declined due to CMOS scaling challenges [FM11], they are still able to outpace any electromechanical efficiency improvements. In the remainder of this chapter we focus on the energy and power efficiency of computing.

5.2.1 MEASURING ENERGY EFFICIENCY

Ultimately, we want to measure the energy consumed to produce a certain result. A number of industry benchmarks try to do exactly that. In high-performance computing (HPC), the Green 500 [TG500] benchmark ranks the energy efficiency of the world's top supercomputers using LIN-PACK. Similarly, server-level benchmarks such as Joulesort [Riv+07] and SPECpower [SPEC] characterize other aspects of computing efficiency. Joulesort measures the total system energy to perform an out-of-core sort and derives a metric that enables the comparison of systems ranging from embedded devices to supercomputers. SPECpower focuses on server-class systems and computes the performance-to-power ratio of a system running a typical business application on an enterprise Java platform. Two separate benchmarking efforts aim to characterize the efficiency of storage systems: the Emerald Program [SNI11] by the Storage Networking Industry Association (SNIA) and the SPC-2/E [SPC12] by the Storage Performance Council. Both benchmarks measure storage servers under different kinds of request activity and report ratios of transaction throughput per watt.

5.2.2 SERVER ENERGY EFFICIENCY

Clearly, the same application binary can consume different amounts of power depending on the server's architecture and, similarly, an application can consume more or less of a server's capacity depending on software performance tuning. Furthemore, systems efficiency can vary with utilization: under low levels of utilization, computing systems tend to be significantly more inefficient than when they are exercised at maximum utilization.

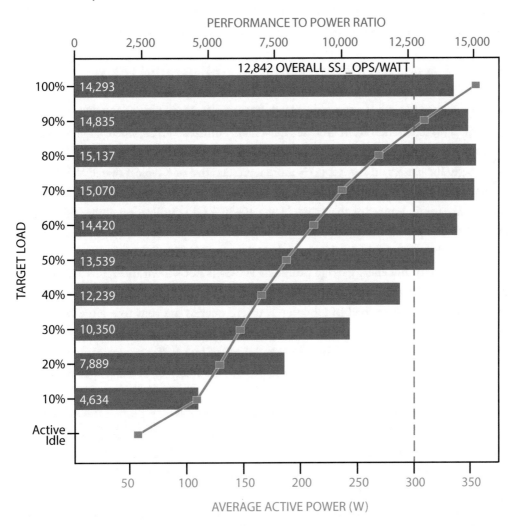

Figure 5.4: Example benchmark result for SPECpower_ssj2008; bars indicate energy efficiency and the line indicates power consumption. Both are plotted for a range of utilization levels, with the average energy efficiency metric corresponding to the vertical dark line. The system has two 2.1 GHz 28-core Intel Xeon processors, 192 GB of DRAM, and one M.2 SATA SSD.

Figure 5.4 shows the SPECpower benchmark results for the top performing entry as of January 2018 under varying utilization. The results show two metrics: performance- (transactions per second) to-power ratio and the average system power, plotted over 11 load levels. One feature in the figure is noteworthy and common to all other SPECpower benchmark results: the performance-to-power ratio drops appreciably as the target load decreases because the system power decreases much more slowly than does performance. Note, for example, that the energy efficiency at 30% load has 30% lower efficiency than at 100%. Moreover, when the system is idle, it is still consuming just under 60 W, which is 16% of the peak power consumption of the server.

5.2.3   USAGE PROFILE OF WAREHOUSE-SCALE COMPUTERS

Figure 5.5 shows the average CPU utilization of two Google clusters during a representative three-month period (measured between January and March 2013); each cluster has over 20,000 servers. The cluster on the right (b) represents one of Google's most highly utilized WSCs, where large continuous batch workloads run. WSCs of this class can be scheduled very efficiently and reach very high utilizations on average. The cluster on the left (a) is more representative of a typical shared WSC, which mixes several types of workloads and includes online services. Such WSCs tend to have relatively low average utilization, spending most of their time in the 10–50% CPU utilization range. This activity profile turns out to be a perfect mismatch with the energy efficiency profile of modern servers in that they spend most of their time in the load region where they are most inefficient.

Another feature of the energy usage profile of WSCs is not shown in Figure 5.5: individual servers in these systems also spend little time idle. Consider, for example, a large web search workload, such as the one described in Chapter 2, where queries are sent to a large number of servers, each of which searches within its local slice of the entire index. When search traffic is high, all servers are being heavily used, but during periods of low traffic, a server might still see hundreds of queries per second, meaning that idle periods are likely to be no longer than a few milliseconds.

The absence of significant idle intervals in general-purpose WSCs, despite the existence of low activity periods, is largely a result of applying sound design principles to high-performance, robust distributed systems software. Large-scale internet services rely on efficient load distribution to a large number of servers, creating a situation such that when load is lighter, we tend to have a lower load in multiple servers instead of concentrating the load in fewer servers and idling the remaining ones. Idleness can be manufactured by the application (or an underlying cluster management system) by migrating workloads and their corresponding state to fewer machines during periods of low activity. This can be relatively easy to accomplish when using simple replication models, when servers are mostly stateless (that is, serving data that resides on a shared NAS or SAN storage system). However, it comes at a cost in terms of software complexity and energy for more complex data distribution models or those with significant state and aggressive exploitation of data locality.

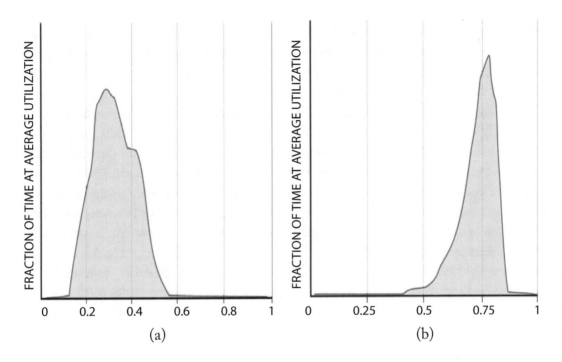

Figure 5.5: Average activity distribution of a sample of 2 Google clusters, each containing over 20,000 servers, over a period of 3 months.

Another reason why it may be difficult to manufacture useful idle periods in large-scale distributed systems is the need for resilient distributed storage. GFS [GGL03] achieves higher resilience by distributing data chunk replicas for a given file across an entire cluster instead of concentrating them within only a small number of machines. This benefits file system performance by achieving fine granularity load balancing, as well as resiliency, because when a storage server crashes (or a disk fails), the replicas in that system can be reconstructed by thousands of machines, making recovery extremely efficient. The consequence of otherwise sound designs is that low traffic levels translate into lower activity for all machines instead of full idleness for a significant subset of them. Several practical considerations may also work against full idleness, as networked servers frequently perform many small background tasks on periodic intervals. The reports on the Tickless kernel project [SPV07] provide other examples of how difficult it is to create and maintain idleness.

5.3   ENERGY-PROPORTIONAL COMPUTING

In an earlier article [BH07], we argued that the mismatch between server workload profile and server energy efficiency behavior must be addressed largely at the hardware level; software alone cannot efficiently exploit hardware systems that are efficient only when they are in inactive idle modes (sleep or standby) or when running at full speed. We believe that systems are inefficient when lightly used largely because of lack of awareness by engineers and researchers about the importance of that region to energy efficiency.

We suggest that *energy proportionality* should be added as a design goal for computing components. Ideally, energy-proportional systems will consume almost no power when idle (particularly in active idle states where they are still available to do work) and gradually consume more power as the activity level increases. A simple way to reason about this ideal curve is to assume linearity between activity and power usage, with no constant factors. Such a linear relationship would make energy efficiency uniform across the activity range, instead of decaying with decreases in activity levels. Note, however, that linearity is not necessarily the optimal relationship for energy savings. As shown in Figure 5.5(a), since servers spend relatively little time at high activity levels, it might be fine to decrease efficiency at high utilizations, particularly when approaching maximum utilization. However, doing so would increase the maximum power draw of the equipment, thus increasing facility costs.

Figure 5.6 illustrates the possible energy efficiency of two hypothetical systems that are more energy-proportional than typical servers. The curves in red correspond to a typical server, circa 2009. The green curves show the normalized power usage and energy efficiency of a more energy-proportional system, which idles at only 10% of peak power and with linear power vs. load behavior. Note how its efficiency curve is far superior to the one for the typical server; although its efficiency still decreases with the load level, it does so much less abruptly and remains at relatively high efficiency levels at 30% of peak load. The curves in blue show a system that also idles at 10% of peak but with a sublinear power versus load relationship in the region of load levels between 0% and 50% of peak load. This system has an efficiency curve that peaks not at 100% load, but around the 30–40% region. From an energy usage standpoint, such behavior would be a good match to the kind of activity spectrum for WSCs depicted in Figure 5.5(a).

The potential gains from energy proportionality in WSCs were evaluated by Fan et al. [FWB07] in their power provisioning study. They used traces of activity levels of thousands of machines over six months to simulate the energy savings gained from using more energy-proportional servers—servers with idle consumption at 10% of peak (similar to the green curves in Figure 5.6) instead of at 50% (such as the corresponding red curve). Their models suggest that energy usage would be halved through increased energy proportionality alone because the two servers compared had the same peak energy efficiency.

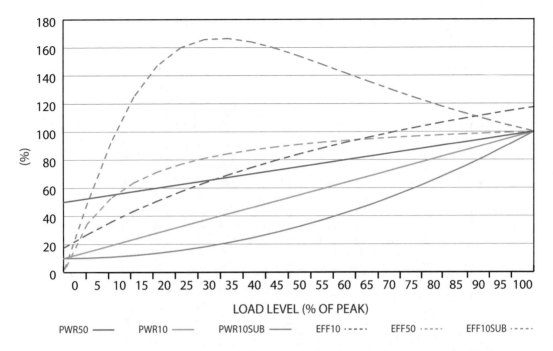

Figure 5.6: Power and corresponding power efficiency of three hypothetical systems: a typical server with idle power at 50% of peak (Pwr50 and Eff50), a more energy-proportional server with idle power at 10% of peak (Pwr10 and Eff10), and a sublinearly energy-proportional server with idle power at 10% of peak (Pwr10sub and Eff10sub). The solid lines represent power % (normalized to peak power). The dashed lines represent efficiency as a percentage of power efficiency at peak.

5.3.1   CAUSES OF POOR ENERGY PROPORTIONALITY

Although CPUs historically have a bad reputation regarding energy usage, they are not necessarily the only culprit for poor energy proportionality. Over the last few years, CPU designers have paid more attention to energy efficiency than their counterparts for other subsystems. The switch to multicore architectures instead of continuing to push for higher clock frequencies and larger levels of speculative execution is one of the reasons for this more power-efficient trend.

The relative contribution of the memory system to overall energy use has decreased over the last five years with respect to CPU energy use, reversing a trend of higher DRAM energy profile throughout the previous decade. The decrease in the fraction of energy used in memory systems is due to a combination of factors: newer DDR3 technology is substantially more efficient than previous technology (FBDIMMs). DRAM chip voltage levels have dropped from 1.8 V to below 1.5 V, new CPU chips use more energy as more aggressive binning processes and temperature-controlled "turbo" modes allow CPUs to run closer to their thermal envelope, and today's systems have a higher ratio of CPU performance per DRAM space (a result of DRAM technology scaling falling behind that of CPUs).

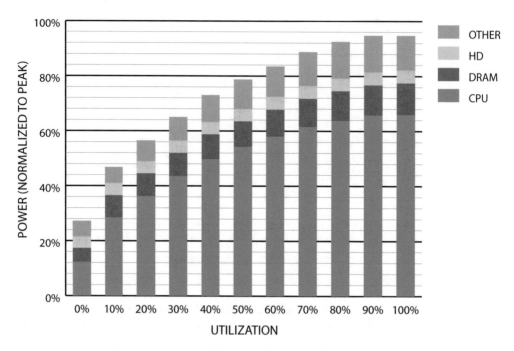

Figure 5.7: Subsystem power usage in an x86 server as the compute load varies from idle to full usage.

Figure 5.7 shows the power usage of the main subsystems for a Google server (circa 2012) as the compute load varies from idle to full activity levels. Unlike what we reported in the first edition, the CPU portion (everything inside a CPU socket) is once more the dominant energy consumer in servers, using two-thirds of the energy at peak utilization and about 40% when (active) idle. In our experience, server-class CPUs have a dynamic power range that is generally greater than 3.0x (more than 3.5x in this case), whereas CPUs targeted at the embedded or mobile markets can do even better. By comparison, the dynamic range of memory systems, disk drives, and networking equipment is much lower: approximately 2.0x for memory, 1.3x for disks, and less than 1.2x for networking switches. This suggests that energy proportionality at the system level cannot be achieved through CPU optimizations alone, but instead requires improvements across all components. Networking and memory are both notable here. Future higher bandwidth memory systems are likely to increase the power of the memory subsystems. Also, given switch radix scaling challenges, the ratio of switches to servers is likely to increase, making networking power more important. Nevertheless, as we'll see later, increased CPU energy proportionality over the last five years, and an increase in the fraction of overall energy use by the CPU, has resulted in more energy proportional servers today.

5.3.2   IMPROVING ENERGY PROPORTIONALITY

Added focus on energy proportionality as a figure of merit in the past five years has resulted in notable improvements for server-class platforms. A meaningful metric of the energy proportionality of a server for a WSC is the ratio between the energy efficiency at 30% and 100% utilizations. A perfectly proportional system will be as efficient at 30% as it is at 100%. In the first edition (in early 2009), that ratio for the top 10 SPECpower results was approximately 0.45, meaning that when used in WSCs, those servers exhibited less than half of their peak efficiency. As of June 2018, that figure has improved almost twofold, reaching 0.80.

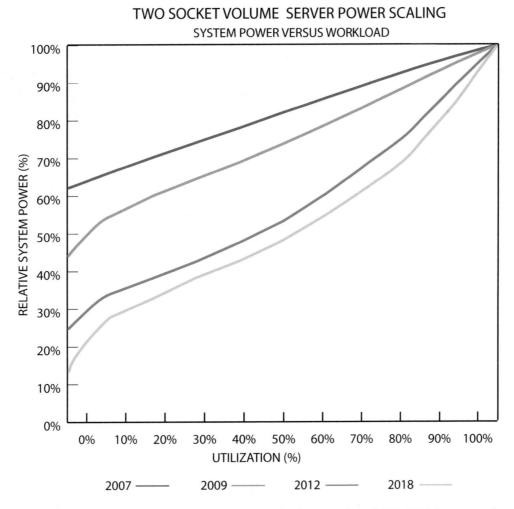

Figure 5.8: Normalized system power vs. utilization in Intel servers from 2007–2018 (courtesy of David Lo, Google). The chart indicates that Intel servers have become more energy proportional in the 12-year period.

Figure 5.8 shows increasing proportionality in Intel reference platforms between 2007 and 2018 [Sou12]. While not yet perfectly proportional, the more recent systems are dramatically more energy proportional than their predecessors.

5.3.3 ENERGY PROPORTIONALITY IN THE REST OF THE SYSTEM

While processor energy proportionality has improved, greater effort is still required for DRAM, storage, and networking. Disk drives, for example, spend a large fraction of their energy budget (as much as 70% of their total power for high RPM drives) simply keeping the platters spinning. Improving energy efficiency and proportionality may require lower rotational speeds, smaller platters, or designs that use multiple independent head assemblies. Carrera et al. [CPB03] considered the energy impact of multi-speed drives and combinations of server-class and laptop drives to achieve proportional energy behavior. Sankar et al. [SGS08] explored different architectures for disk drives, observing that because head movements are relatively energy-proportional, a disk with lower rotational speed and multiple heads might achieve similar performance and lower power when compared with a single-head, high RPM drive.

Traditionally, data center networking equipment has exhibited rather poor energy proportionality. At Google we have measured switches that show little variability in energy consumption between idle and full utilization modes. Historically, servers didn't need much network bandwidth, and switches were expensive, so their overall energy footprint was relatively small (in the single digit percentages of total IT power). However, as switches become more commoditized and bandwidth needs increase, networking equipment could become responsible for 10–20% of the facility energy budget. At that point, their lack of proportionality will be a severe problem. To illustrate this point, let's assume a system that exhibits a linear power usage profile as a function of utilization (u):

$$P(u) = Pi + u(1\text{-}Pi).$$

In the equation above, Pi represents the system's idle power, and peak power is normalized to 1.0. In such a system, energy efficiency can be estimated as u/P(u), which reduces to the familiar Amdahl Law formulation below:

$$E(u) = \frac{1}{1 - Pi + Pi/u}.$$

Unlike the original Amdahl formula, we are not interested in very high values of u, since it can only reach 1.0. Instead, we are interested in values of utilization between 0.1 and 0.5. In that case, high values for Pi (low energy proportionality) will result in low efficiency. If every subcomponent of a WSC is highly energy proportional except for one (say, networking or storage), that subcomponent will limit the whole system efficiency similarly to how the amount of serial work limits parallel speedup in Amdahl's formula.

Efficiency and proportionality of data center networks might improve in a few ways. Abts et al. [Abt+10] describe how modern plesiochronous links can be modulated to adapt to usage as well as how topology changes and dynamic routing can create more proportional fabrics. The IEEE's Energy-Efficient Ethernet standardization effort [Chr+10], (802.3az), is also trying to pursue interoperable mechanisms that allow link-level adaptation.

Finally, energy-proportional behavior is not only a target for electronic components but for the entire WSC system, including power distribution and cooling infrastructure.

5.3.4   RELATIVE EFFECTIVENESS OF LOW-POWER MODES

As discussed earlier, long idleness intervals would make it possible to achieve higher energy proportionality by using various kinds of sleep modes. We call these low-power modes *inactive* because the devices are not usable while in those modes, and typically a sizable latency and energy penalty is incurred when load is reapplied. Inactive low-power modes were originally developed for mobile and embedded devices, and they are very successful in that domain. However, most of those techniques are a poor fit for WSC systems, which would pay an inactive-to-active latency and energy penalty too frequently. The few techniques that can be successful in this domain have very low wake-up latencies, as is beginning to be the case with CPU low-power halt states (such as the ACPI C1E state).

Unfortunately, these tend to be the low-power modes with the smallest degrees of energy savings. Large energy savings are available from inactive low-power modes such as spun-down disk drives. A spun-down disk might use almost no energy, but a transition to active mode incurs a latency penalty 1,000 times higher than a regular access. Spinning up the disk platters adds an even larger energy penalty. Such a huge activation penalty restricts spin-down modes to situations in which the device will be idle for several minutes, which rarely occurs in servers.

Active low-power modes save energy at a performance cost while not requiring inactivity. CPU voltage-frequency scaling is an example of an active low-power mode because it remains able to execute instructions albeit at a slower rate. The (presently unavailable) ability to read and write to disk drives at lower rotational speeds is another example of this class of low-power modes. In contrast with inactive modes, active modes are useful even when the latency and energy penalties to transition to a high-performance mode are significant. Because active modes are operational, systems can remain in low-energy states for as long as they remain below certain load thresholds. Given that periods of low activity are more common and longer than periods of full idleness, the overheads of transitioning between active energy savings modes amortize more effectively.

The use of very low-power inactive modes with high-frequency transitions has been proposed by Meisner et al. [MGW09] and Gandhi et al. [Gan+] as a way to achieve energy proportionality. The systems proposed, PowerNap and IdleCap, assume that subcomponents have no useful low power modes other than full idleness and modulate active-to-idle transitions in all subcomponents in order to reduce power at lower utilizations while limiting the impact on performance. The promise of such approaches hinges on system-wide availability of very low power idle modes with very short active-to-idle and idle-to-active transition times, a feature that seems within reach for processors but more difficult to accomplish for other system components. In fact, Meisner et al. [Mei+11] analyze the behavior of online data intensive workloads (such as web search) and conclude that existing low power modes are insufficient to yield both energy proportionality and low latency.

5.3.5 THE ROLE OF SOFTWARE IN ENERGY PROPORTIONALITY

We have argued that hardware components must undergo significant improvements in energy proportionality to enable more energy-efficient WSC systems. However, more intelligent power management and scheduling software infrastructure plays an important role too. For some component types, achieving perfect energy-proportional behavior may not be a realizable goal. Designers will have to implement software strategies for intelligent use of power management features in existing hardware, using low-overhead inactive or active low-power modes, as well as implementing power-friendly scheduling of tasks to enhance energy proportionality of hardware systems. For example, if the activation penalties in inactive low-power modes can be made small enough, techniques like PowerNap (Meisner et al. [MGW09]) could be used to achieve energy-proportional behavior with components that support only inactive low-power modes.

This software layer must overcome two key challenges: encapsulation and performance robustness. Energy-aware mechanisms must be encapsulated in lower-level modules to minimize exposing additional infrastructure complexity to application developers; after all, WSC application developers already deal with unprecedented scale and platform-level complexity. In large-scale systems, completion of an end-user task also tends to depend on large numbers of systems performing at adequate levels. If individual servers begin to exhibit excessive response time variability as a result of mechanisms for power management, the potential for service-level impact is fairly high and can lead to the service requiring additional machine resources, resulting in minimal net improvements.

Incorporating end-to-end metrics and service level objective (SLO) targets from WSC applications into power-saving decisions can greatly help overcome performance variability challenges while moving the needle toward energy proportionality. During periods of low utilization, latency slack exists between the (higher latency) SLO targets and the currently achieved latency. This slack represents power saving opportunities, as the application is running faster than needed. Having end-to-end performance metrics is a critical piece needed to safely reduce the performance of the WSC in response to lower loads. Lo et al. [Lo+14] propose and study a system (PEGA-SUS) that combines hardware power actuation mechanisms (Intel RAPL [Intel18]) with software control policies. The system uses end-to-end latency metrics to drive decisions on when to adjust CPU power in response to load shifts. By combining application-level metrics with fine-grained hardware actuation mechanisms, the system is able to make overall server power more energy proportional while respecting the latency SLOs of the WSC application.

Software plays an important role in improving cluster-level energy efficiency despite poor energy proportionality of underlying servers. By increasing the utilization of each individual server, cluster management software can avoid operating servers in the region of poor energy efficiency at low loads. Cluster scheduling software such as Borg [Ver+15] and Mesos [Hin+11] take advantage of resource sharing to significantly improve machine-level utilization through better bin-packing of disparate jobs (encapsulation). This is a net win for energy efficiency, where the poor energy proportionality of the servers that make up a WSC is mitigated by running the server at higher utilizations closer to its most energy efficient operating point. An even larger benefit of higher utilization is that the number of servers needed to serve a given capacity requirement is reduced, which lowers the TCO dramatically due to a significant portion of the cost of a WSC being in concentrated in the CapEx costs of the hardware.

However, as server utilization is pushed higher and higher, performance degradation from shared resource contention becomes a bigger and bigger issue. For example, if two workloads that would each completely saturate DRAM bandwidth are co-located on the same server, then both workloads will suffer significantly degraded performance compared to when each workload is run in isolation. With workload agnostic scheduling, the probability of this scenario occurring increases as server capacity increases with the scaling of CPU core counts. To counter the effects of interference, service owners tend to increase the resource requirements of sensitive workloads in order to ensure that their jobs will have sufficient compute capacity in the face of resource contention. This extra padding has an effect of lowering server utilization, thus also negatively impacting energy efficiency. To avoid this pitfall and to further raise utilization, contention aware scheduling needs to be utilized. Systems such as Bubble-Up [Mar+11], Heracles [Lo+15], and Quasar [DK14] achieve significantly higher server utilizations while maintaining strict application-level performance performance requirements. While the specific mechanisms differ for each system, they all share a common trait of using performance metrics in making scheduling and resource allocation decisions to provide both encapsulation and performance robustness for workloads running in the WSC. By overcoming these key challenges, such performance-aware systems can lead to significantly more resource sharing opportunities, increased machine utilization, and ultimately energy efficient WSCs that can sidestep poor energy proportionality.

Raghavendra et al. [Rag+08] studied a five-level coordinated power management scheme, considering per-server average power consumption, power capping at the server, enclosure, and group levels, as well as employing a virtual machine controller (VMC) to reduce the average power consumed across a collection of machines by consolidating workloads and turning off unused machines. Such intensive power management poses nontrivial control problems. For one, applications may become unstable if some servers unpredictably slow down due to power capping. On the implementation side, power capping decisions may have to be implemented within milliseconds to avoid tripping a breaker. In contrast, overtaxing the cooling system may result in "only" a temporary thermal excursion, which may not interrupt the performance of the WSC. Nevertheless, as individual servers consume more power with a larger dynamic range due to improving energy proportionality in hardware, power capping becomes more attractive as a means of fully realizing the compute capabilities of a WSC.

Wu et al. [Wu+16a] proposed and studied the use of Dynamo, a dynamic power capping system in production at Facebook. Dynamo makes coordinated power decisions across the entire data center to safely oversubscribe power and improve power utilization. Using Intel RAPL as the node-level enforcement mechanism to cap machine power, the system is workload-aware to ensure that high priority latency-sensitive workloads are throttled only as a measure of last resort. As a result of deploying Dynamo, the authors note a significant boost in power capacity utilization at their data centers through increased use of dynamic core frequency boosting; namely, Intel Turbo Boost [IntTu], which can run CPU cores at higher frequencies given sufficient electrical and thermal headroom. Much like PEGASUS, Dynamo combines application-specific knowledge with fine-grained hardware knobs to improve the realizable compute capability of the WSC while respecting application performance boundaries.

Technologies such as Turbo Boost reflect a growing trend in CPU design of adding additional dynamic dimensions (CPU activity) to trade off power and performance. The behavior of Turbo Boost is highly dependent on the number of active cores and the compute intensity of the workload. For example, CPU core frequency can vary by as much as 85% on Intel Skylake server CPUs [IntXe]. Another manifestation of this phenomenon takes the form of wider vector instructions, such as AVX-512, which can cause large drops in CPU core frequency due to its usage. On the one hand, these techniques enable higher peak performance, but on the other hand, they increase performance variability across the WSC. Dynamic frequency scaling decisions made in hardware present a set of new challenges in achieving performance robustness, and software designers must be cognizant of such effects in hardware and to handle the resulting performance variation.

5.4   ENERGY EFFICIENCY THROUGH SPECIALIZATION

So far we have assumed traditional WSCs: collections of servers, each with CPUs, DRAM, networking, and disks; all computation handled by general purpose CPUs. However, recapping the discussion in Chapter 4, Dennard scaling has now ended (due to fundamental device limitations that prevent operating voltage from further being scaled due to static leakage concerns), and Moore's Law is well on its way to being sunset (as chip manufacturers struggle with maintaining high yield while further shrinking transistor sizes). Looking forward, general purpose CPUs are facing a daunting task when it comes to further energy efficiency improvements. This issue is orthogonal to energy proportionality, as it is about improving energy efficiency at peak compute load.

While general-purpose CPUs improve marginally over time when it comes to energy efficiency improvements at peak load, the demand for compute is growing at a steady rate. Currently, this demand is being driven by technologies powered by artificial intelligence and machine learning, which require extraordinary amounts of compute commensurate with large model sizes and gargantuan amounts of data being fed into such workloads. While general-purpose CPUs are fully capable of performing the operations necessary for artificial intelligence, they are not optimized to run these kinds of workloads.

Specialized accelerators are designed for running one particular class of workloads well. The hardware for these accelerators can be general purpose graphics processing units (GPGPUs), field programmable gate arrays (FPGAs), and application-specific integrated circuits (ASICs), to name a few. Unlike general-purpose CPUs, specialized accelerators are incapable of running all kinds of workloads with reasonable efficiency. That is because these accelerators trade off general compute capabilities for the ability to run a subset of workloads with phenomenal performance and efficiency. High-performance server class CPUs are designed to extract the maximum performance out of challenging workloads with a wide variety of potential kinds of computations, unpredictable control flows, irregular to non-existent parallelism, and complicated data dependencies. On the other hand, specialized accelerators need to perform well only for a specific kind of computation that provides opportunities for domain-specific optimizations.

For example, consider Google's Tensor Processing Unit (TPU) [Jou+17]. This custom ASIC was designed to handle the inference portion of several machine learning workloads. The energy efficiency of the TPU benefited greatly from the specialization of compute. The TPU is powered by a systolic array, an energy efficient construct that excels at performing regular computations, such as matrix multiplication. The use of a systolic array allows the TPU to avoid a high access rate to large SRAM arrays that would otherwise consume significant amounts of power. In addition, compared to a modern superscalar out-of-order CPU, the control logic for a TPU is relatively simple and thus much more energy efficient. Since parallelism in machine learning applications is easier to extract, the TPU has no need for the complicated and energy hungry control hardware found in CPUs. These and other design decisions for the TPU unlocked a vast improvement in energy efficiency.

Figure 5.9 shows how the TPU is orders of magnitude more energy efficient for inference tasks compared to a contemporary server CPU of its time (Intel Haswell). However, while the energy efficiency is high, the energy proportionality of the TPU happens to be much worse than that of the CPU, as it consumes 88% of peak power at 10% load. The designers note that the poor energy proportionality is not due to fundamental reasons but to tradeoffs around design expediency. Nevertheless, there is an open opportunity to apply the same learnings from general compute, such as improved energy proportionality, to specialized accelerators as well.

Specialized accelerators have an important role to play in improving the energy efficiency of WSCs of the future. Large emerging workloads such as machine learning are ripe targets for acceleration due to the sheer volume of compute they demand. The challenge is to identify workloads that benefit from being implemented on specialized accelerators and to progress from concept to product in a relatively short timespan. In addition, the same insights from improving energy efficiency of servers (such as energy proportionality) also apply to accelerators. Nevertheless, not all workloads can be put on specialized compute hardware. These can be due to the nature of the workload itself (general-purpose CPUs can be viewed as accelerators for complex, branchy, and irregular code) or due to it not having enough deployment volume to justify the investment in specialized hardware. Thus, it is still important to improve the overall energy efficiency of the entire data center, general-purpose servers included, in conjunction with improving energy efficiency of accelerators.

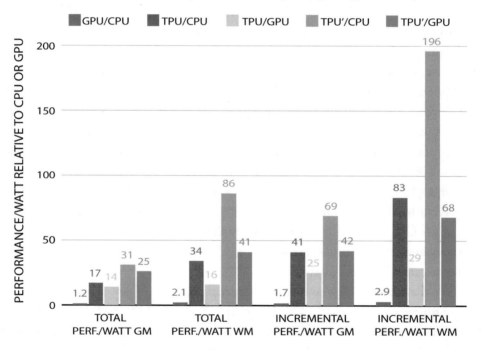

Figure 5.9: Relative performance/watt (TDP) of GPU server (blue bar) and TPU server (red bar) to CPU server, and TPU server to GPU server (orange bar). TPU' is an improved TPU. The green bar shows its ratio to the CPU server and the blue bar shows its relation to the GPU server. Total includes host server power, but incremental doesn't. GM and WM are the geometric and weighted means [Jou+17].

5.5   DATA CENTER POWER PROVISIONING

Energy efficiency optimizations reduce electricity costs. In addition, they reduce construction costs. For example, if free cooling eliminates the need for chillers, then we don't have to purchase and install those chillers, nor do we have to pay for generators or UPSs to back them up. Such construction cost savings can double the overall savings from efficiency improvements.

Actually *using* the provisioned power of a facility is equally important. For example, if a facility operates at 50% of its peak power capacity, the effective provisioning cost per watt used is doubled. This incentive to fully use the power budget of a data center is offset by the risk of exceeding its maximum capacity, which could result in outages.

5.5.1   DEPLOYING THE RIGHT AMOUNT OF EQUIPMENT

How many servers can we install in a 1 MW facility? This simple question is harder to answer than it seems. First, server specifications usually provide very conservative values for maximum power consumption. Some vendors, such as Dell and HP, offer online power calculators [DEC, HPPC] to provide better estimates, but it may be necessary to measure the actual power consumption of the dominant applications manually.

Second, actual power consumption varies significantly with load (thanks to energy proportionality), and it may be hard to predict the peak power consumption of a group of servers. While any particular server might temporarily run at 100% utilization, the maximum utilization of a group of servers probably isn't 100%. But to do better, we'd need to understand the correlation between the simultaneous power usage of large groups of servers. The larger the group of servers and the higher the application diversity, the less likely it is to find periods of simultaneous very high activity.

5.5.2   OVERSUBSCRIBING FACILITY POWER

As soon as we use anything but the most conservative estimate of equipment power consumption to deploy clusters, we incur a certain risk that we'll exceed the available amount of power; that is, we'll *oversubscribe* facility power. A successful implementation of power oversubscription increases the overall utilization of the data center's power budget while minimizing the risk of overload situations. We will expand on this issue because it has received much less attention in technical publications than the first two steps listed above, and it is a very real problem in practice [Man09].

Fan et al. [FWB07] studied the potential opportunity of oversubscribing facility power by analyzing power usage behavior of clusters with up to 5,000 servers running various workloads at Google during a period of six months. One of their key results is summarized in Figure 5.10, which shows the cumulative distribution of power usage over time for groups of 80 servers (Rack), 800 servers (PDU), and 5,000 servers (Cluster).

Power is normalized to the peak aggregate power of the corresponding group. For example, the figure shows that although rack units spend about 80% of their time using less than 65% of their peak power, they do reach 93% of their peak power at some point during the six month observation window. For power provisioning, this indicates a very low oversubscription opportunity at the rack level because only 7% of the power available to the rack was stranded. However, with larger machine groups, the situation changes. In particular, the whole cluster never ran above 72% of its aggregate peak power. Thus, if we had allocated a power capacity to the cluster that corresponded to the *sum of the peak* power consumption of all machines, 28% of that power would have been stranded. This means that within that power capacity, we could have hosted nearly 40% more machines.

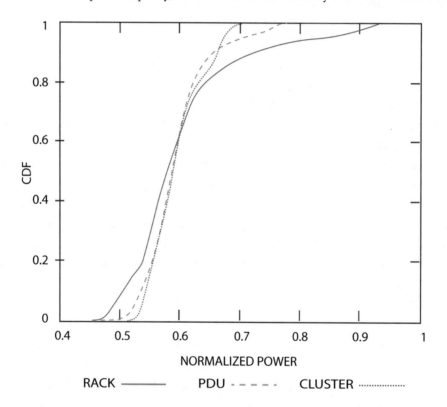

Figure 5.10: Cumulative distribution of time that groups of machines spend at or below a given power level (power level is normalized to the maximum peak aggregate power for the corresponding grouping) (Fan et al. [FWB07]).

This study also evaluates the potential of more energy-proportional machines to reduce peak power consumption at the facility level. It suggests that lowering idle power from 50% to 10% of peak (that is, going from the red to the green curve in Figure 5.6) can further reduce cluster peak power usage by more than 30%. This would be equivalent to an additional 40%+ increase in facility hosting capacity.

The study further found that mixing different workloads within a cluster increased the opportunities for power oversubscription because this reduces the likelihood of synchronized power peaks across machines. Once oversubscription is applied, the system needs a safety mechanism to handle the possibility that workload changes may cause the power draw to exceed the data center capacity. This can be accomplished by always allocating some fraction of the computing resources to a workload that runs in a lower priority class or that otherwise does not have strict deadlines to meet (many batch workloads may fall into that category). Such workloads can be quickly paused or aborted to reduce facility load. Provisioning should not be so aggressive as to require this mechanism to be triggered often, which might be the case if oversubscription is applied at the rack level, for example.

In a real deployment, it's easy to end up with an underutilized facility even when you pay attention to correct power ratings. For example, a facility typically needs to accommodate future growth, but keeping space open for such growth reduces utilization and thus increases unit costs. Various forms of fragmentation can also prevent full utilization. Perhaps we run out of space in a rack because low-density equipment used it up, or we can't insert another server because we're out of network ports, or we're out of plugs or amps on the power strip. For example, a 2.5 kW circuit supports only four 520 W servers, limiting utilization to 83% on that circuit. Since the lifetimes of various WSC components differ (servers might be replaced every 3 years, cooling every 10 years, networking every 4 years, and so on) it's difficult to plan for 100% utilization, and most organizations don't.

Management of energy, peak power, and temperature of WSCs are becoming the targets of an increasing number of research studies. Chase et al. [Cha+01c], G. Chen et al. [Che+07], and Y. Chen et al. [Che+05] consider schemes for automatically provisioning resources in data centers, taking energy savings and application performance into account. Raghavendra et al. [Rag+08] describe a comprehensive framework for power management in data centers that coordinates hardware-level power capping with virtual machine dispatching mechanisms through the use of a control theory approach. Femal and Freeh [FF04, FF05] focus specifically on the issue of data center power oversubscription and describe dynamic voltage-frequency scaling as the mechanism to reduce peak power consumption. Managing temperature is the subject of the systems proposed by Heath et al. [Hea+06] and Moore et al. [Moo+05]. Finally, Pedram [Ped12] provides an introduction to resource provisioning and summarizes key techniques for dealing with management problems in the data center. Incorporating application-level knowledge to safely save power by re-shaping its latency distribution through DVFS is studied by Lo et al. [Lo+14], Kasture et al. [Kas+15], and Hsu et al. [Hsu+15].

5.6   TRENDS IN SERVER ENERGY USAGE

While in the past dynamic voltage and frequency scaling (DVFS) was the predominant mechanism for managing energy usage in servers, today we face a different and more complex scenario. Given lithography scaling challenges, the operating voltage range of server-class CPUs is very narrow, resulting in ever decreasing gains from DVFS.

Figure 5.11 shows the potential power savings of CPU dynamic voltage scaling (DVS) for the same server by plotting the power usage across a varying compute load for three frequency-voltage steps. Savings of approximately 10% are possible once the compute load is less than two thirds of peak by dropping to a frequency of 1.8 GHz (above that load level the application violates latency SLAs). An additional 10% savings is available when utilization drops to one third by going to a frequency of 1 GHz. However, as the load continues to decline, the gains of DVS once again return to a maximum of 10%. Instead, modern CPUs increasingly rely on multiple power planes within a die as their primary power management mechanism, allowing whole sections of the chip to be powered down and back up quickly as needed. As the number of CPUs in a die increases, such coarse-grained power gating techniques will gain greater potential to create energy proportional systems.

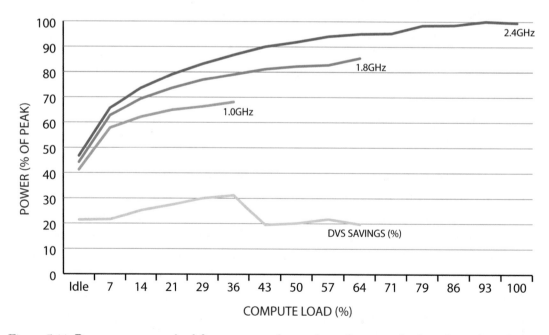

Figure 5.11: Power vs. compute load for a server at three voltage-frequency levels and corresponding energy savings.

A second trend is that CPUs continue to outpace other server components in energy proportionality improvements. The result is a power budget breakdown with larger energy fractions from non-CPU subsystems at *lower* utilizations.

5.6.1   USING ENERGY STORAGE FOR POWER MANAGEMENT

Several studies [Gov+, Wan+, Kon+12] propose using energy stored in the facility's backup systems (such as UPS batteries) to optimize facility performance or reduce energy costs. Stored energy could be used to flatten the facility's load profile (using less utility power when it's most expensive), mitigate supply variability in a wind-powered facility, or manage short demand peaks in oversubscribed facilities (using stored energy instead of capping the load).

In our opinion, the most promising use of energy storage in power management consists of managing short demand peaks or short-term supply reductions (say, when a data center is partly powered by renewable energy sources, such as wind). Power capping systems need some time to react intelligently to demand peak events, and may need to set peak provisioning levels well below the maximum breaker capacity in order to allow time for power capping to respond. A power capping system that can draw from energy storage sources for just a few seconds during an unexpected peak would allow the facility to safely operate closer to its maximum capacity while requiring a relatively modest amount of additional energy storage capacity.

To our knowledge no such power management systems have yet been used in production systems. Deploying such a system would be difficult and potentially costly. Besides the control complexity, the additional cost of batteries would be significant, since we couldn't just reuse the existing UPS capacity for power management, as doing so would make the facility more vulnerable in an outage. Furthermore, the types of batteries typically used in UPS systems (lead-acid) don't age well under frequent cycling, so that more expensive technologies might be required. While some have argued that expanded UPSs would be cost effective [Kon+12], we believe that the economic case has not yet been made in practice.

5.7   SUMMARY

Energy efficiency is a key cost driver for WSCs, and we expect energy usage to become an increasingly important factor in WSC design. The current state of the industry is poor: the average real-world data center and the average server are far too inefficient, mostly because efficiency has historically been neglected and has taken a backseat relative to reliability, performance, and capital expenditures. As a result, the average WSC wastes two thirds or more of its energy.

The upside of this history of neglect is that sizable improvements are almost trivial to obtain—an overall factor of two in efficiency improvements is possible, without much risk, by simply applying best practices to data center and server designs. Unfortunately, the path beyond this low-hanging fruit is more difficult, posing substantial challenges to overcome inherently complex problems and often unfavorable technology trends. Once the average, data center achieves state-of-the-art PUE levels, and servers are deployed with high-efficiency power supplies that are available today, the opportunity for further efficiency improvements in those areas drops to below 40%. From a research and development standpoint, greater opportunities for gains in energy efficiency from now on will need to come from computer scientists and engineers, and less so from mechanical or power conversion specialists (though large opportunities remain for mechanical and power engineers in reducing facility costs in particular).

First, power and energy must be better managed to minimize operational cost. Power determines overall facility cost because much of the construction cost is directly related to the maximum power draw that must be supported. Overall energy usage determines the electricity bill as well as much of the environmental impact. Today's servers can have high maximum power draws that are rarely reached in practice, but that must be accommodated or limited to avoid overloading the facility's power delivery system. Power capping promises to manage the aggregate power of a pool of servers, but it is difficult to reconcile with availability; that is, the need to use peak processing power in an emergency caused by a sudden spike in traffic or by a failure in another data center. In addition, peak server power is increasing despite the continuing shrinking of silicon gate sizes, driven by a combination of increasing operating frequencies, larger cache and memory sizes, and faster off-chip communication (DRAM and I/O buses as well as networking speeds).

Second, today's hardware does not gracefully adapt its power usage to changing load conditions, and as a result, a server's efficiency degrades seriously under light load. Energy proportionality promises a way out of this dilemma but may be challenging to implement across all subsystems. For example, disks do not naturally lend themselves to lower-power active states. Systems for work consolidation that free up and power down entire servers present an avenue to create energy-proportional behavior in clusters built with non-energy-proportional components but are harder to implement and manage, requiring transparent process migration and degrading the WSC's ability to react to sudden upticks in load. Furthermore, high-performance and high-availability distributed systems software tends to spread data and computation in a way that reduces the availability of sufficiently large idle periods on any one system. Energy-management-aware software layers must then manufacture idleness in a way that minimizes the impact on performance and availability.

Third, energy optimization is a complex end-to-end problem, requiring intricate coordination across hardware, operating systems, VMs, middleware, applications, and operations organizations. Even small mistakes can ruin energy savings; for example, when a suboptimal device driver generates too many interrupts or when network chatter from neighboring machines keeps a machine from quiescing. There are too many components involved for perfect coordination to happen naturally, and we currently lack the right abstractions to manage this complexity. In contrast to hardware improvements, such as energy-proportional components that can be developed in relative isolation, solving this end-to-end problem at scale will be much more difficult.

Fourth, the hardware performing the computation can be made more energy efficient. General purpose CPUs are generally efficient for any kind of computation, which is to say that they are not super efficient for any particular computation. ASICs and FPGAs trade off generalizability for better performance and energy efficiency. Special-purpose accelerators (such as Google's tensor processing units) are able to achieve orders of magnitude better energy efficiency compared to general purpose processors. With the sunset of Moore's Law and the breakdown of Dennard scaling, specializing compute will likely keep its place as one of the remaining tools in the shrinking toolbox of hardware changes that can further improve energy efficiency.

Finally, this discussion of energy optimization shouldn't distract us from focusing on improving server utilization, since that is the best way to improve cost efficiency. Underutilized machines aren't only inefficient per unit of work, they're also expensive. After all, you paid for all those servers, so you'd better keep them doing something useful. Better resource sharing through cluster-level scheduling and performance-aware scheduling have made very promising forward progress in increasing server utilization while maintaining workload encapsulation and performance robustness.

CHAPTER 6

Modeling Costs

As described in Chapter 1, one of the defining characteristics of WSCs is their emphasis on cost efficiency at scale. To better understand this, let us examine the total cost of ownership (TCO) of a data center. At the top level, costs split into capital expenses (Capex) and operational expenses (Opex). Capex refers to investments that must be made upfront and that are then depreciated over a certain timeframe. Examples are the construction cost of a data center or the purchase price of a server. Opex refers to the recurring monthly costs of actually running the equipment, excluding depreciation: electricity costs, repairs and maintenance, salaries of on-site personnel, and so on. Thus, we have:

TCO = data center depreciation + data center Opex + server depreciation + server Opex.

We focus on top-line estimates in this chapter, simplifying the models where appropriate. More detailed cost models can be found in the literature [Pat+05, Koo+07]. For academic purposes, our simplified model is accurate enough to model all major costs; the primary source of inaccuracy compared to real-world data centers will be the model input values, such as the cost of construction.

6.1 CAPITAL COSTS

Data center construction costs vary widely depending on design, size, location, and desired speed of construction. Not surprisingly, adding reliability and redundancy makes data centers more expensive, and very small or very large data centers tend to be more expensive (the former because fixed costs cannot be amortized over many watts, the latter because large data centers require additional infrastructure, such as electrical substations).

Table 6.1 shows a range of typical data center construction costs, expressed in dollars per watt of usable critical power, drawn from a variety of sources. In general, most large enterprise data centers cost around $9–13 per watt to build, and smaller ones cost more. The cost numbers in the table below shouldn't be directly compared since the scope of the projects may differ. For example, the amount quoted may or may not include land or the cost of a pre-existing building.

Cost/W	Source
$12-25	Uptime Institute estimates for small- to medium-sized data centers; the lower value is for Tier I designs that are rarely built in practice [TS06].
$9-13	Dupont Fabros 2011 Form 10K report [DuP11] contains financial information suggesting the following cost for its most recent facilities (built in 2010 and 2011; see page 39 for critical load and page 76 for cost): $204M for 18.2 MW (NJ1 Phase I) => $11.23/W $116M for 13 MW (ACC6 Phase I) => $8.94/W $229M for 18.2 MW (SC1 Phase 1) => $12.56/W
$8-10	Microsoft's investment of $130M for 13.2 MW ($9.85/W) capacity expansion to its data center in Dublin, Ireland [Mic12]. Facebook is reported to have spent $210M for 28 MW ($7.50/W) at its Prineville data center [Mil12].

Table 6.1: Range of data center construction costs expressed in U.S. dollars per watt of critical power. Critical power is defined as the peak power level that can be provisioned to IT equipment

Historical costs of data center construction of Tier III facilities range from $9–$13 per watt. The recent growth of cloud computing is driving a data center construction boom. North American Data Center reports[5] nearly 300 MW under construction in 2017, a 5-year high. As data center construction projects continue to increase, costs are falling. The costs of most recent constructions range from $7–$9 per watt, as revealed in Form 10K reports from companies including Digital Realty Trust (DLR), CyrusOne (CONE), and QTS Realty Trust (QTS) [GDCC].

Characterizing cost in terms of dollars per watt makes sense for larger data centers (where size-independent fixed costs are a relatively small fraction of overall cost) because all of the data center's primary components—power, cooling, and space—roughly scale linearly with watts. Typically, approximately 60–80% of total construction cost goes toward power and cooling, and the remaining 20–40% toward the general building and site construction.

Cost varies with the degree of desired redundancy and availability, and thus we always express cost in terms of *dollars per critical watt*; that is, watts that can actually be used by IT equipment. For example, a data center with 20 MW of generators may have been built in a 2N configuration and provide only 6 MW of critical power (plus 4 MW to power chillers). Thus, if construction costs $120 million, it costs $20/W, not $6/W. Industry reports often do not correctly use the term critical power, so our example data center might be described as a 20 MW data center or even a 30 MW data center if it is supplied by an electrical substation that can provide 30 MW.

Frequently, construction cost is quoted in dollars per square foot, but that metric is less useful because it cannot adequately compare projects and is used even more inconsistently than cost

[5]   https://nadatacenters.com/wp-content/uploads/NADC-Newsletter-2018-R4.pdf

expressed in dollars per watt. In particular, there is no standard definition of what space to include or exclude in the computation, and the metric does not correlate well with the primary cost driver of data center construction, namely critical power. Thus, most industry experts avoid using dollars per square foot to express cost.

The monthly depreciation cost (or amortization cost) that results from the initial construction expense depends on the duration over which the investment is amortized (related to its expected lifetime) and the assumed interest rate. Typically, data centers are depreciated over periods of 15–20 years. Under U.S. accounting rules, it is common to use straight-line depreciation where the value of the asset declines by a fixed amount each month. For example, if we depreciate a $12/W data center over 12 years, the depreciation cost is $0.08/W per month. If we took out a loan to finance construction at an interest rate of 8%, the associated monthly interest payments add an additional $0.05/W, for a total of $0.13/W per month. Typical interest rates vary over time, but many companies use a cost of capital rate in the 7–12% range.

Server costs are computed similarly, except that servers have a shorter lifetime and thus are typically depreciated over 3–4 years. To normalize server and data center costs, it is useful to characterize server costs per watt as well, using the server's peak real-life power consumption as the denominator. For example, a $4,000 server with an actual peak power consumption of 500 W costs $8/W. Depreciated over 4 years, the server costs $0.17/W per month. Financing that server at 8% annual interest adds another $0.02/W per month, for a total of $0.19/W per month.

As discussed in earlier chapters, with the slowdown of Moore's law, WSCs are increasingly turning to hardware accelerators to improve performance per watt. The Capex of internally developed accelerators also includes non-recurring engineering (NRE), the cost of designing and fabricating the ASIC, as well the surrounding infrastructure. If developing and deploying 100,000 accelerators has a one time cost of $50M with each accelerator consuming 200 W, depreciated over 4 years, the NRE cost is $0.05/W per month.

6.2 OPERATIONAL COSTS

Data center operational expense (Opex) is harder to characterize because it depends heavily on operational standards (for example, how many security guards are on duty at the same time or how often generators are tested and serviced) as well as on the data center's size: larger data centers are cheaper because fixed costs are amortized better. Costs can also vary depending on geographic location (climate, taxes, salary levels, and so on) and on the data center's design and age. For simplicity, we will break operational cost into a monthly charge per watt that represents items like security guards and maintenance, and electricity. Typical operational costs for multi-megawatt data centers in the U.S. range from $0.02–$0.08/W per month, excluding the actual electricity costs.

Similarly, servers have an operational cost. Because we are focusing just on the cost of running the infrastructure itself, we will focus on just hardware maintenance and repairs as well as electricity costs. Server maintenance costs vary greatly depending on server type and maintenance standards (for example, response times for four hours vs. two business days).

Also, in traditional IT environments, the bulk of the operational cost lies in the applications; that is, software licenses and the cost of system administrators, database administrators, network engineers, and so on. We are excluding these costs here because we are focusing on the cost of running the physical infrastructure, but also because application costs vary greatly depending on the situation. In small corporate environments, it is not unusual to see one system administrator per a few tens of servers, resulting in a substantial per-machine annual cost [RFG02]. Many published studies attempt to quantify administration costs, but most of them are financed by vendors trying to prove the cost-effectiveness of their products, so that reliable unbiased information is scarce. However, it is commonly assumed that large-scale applications require less administration, scaling to perhaps 1,000 servers per administrator.

6.3   CASE STUDIES

Given the large number of variables involved, it is best to illustrate the range of cost factors by looking at a small number of case studies that represent different kinds of deployments.

First, we consider a typical new multi-megawatt data center in the U.S. (something closer to the Uptime Institute's Tier III classification), fully populated with servers at the high end of what can still be considered a volume rack-mountable server product. For this example we chose a Dell PowerEdge FC640, with 2 CPUs, 128 GB of RAM, and 960 GB SSD. This server draws 340 W at peak per Dell's configuration planning tool and costs approximately $5,000 as of 2018. The remaining base case parameters chosen are as follows.

- The cost of electricity is the 2018 average U.S. industrial rate of 6.7 cents/kWh.

- The interest rate a business must pay on loans is 8%, and we finance the servers with a 3-year interest-only loan.

- The cost of data center construction is $10/W, amortized over 20^6 years.

- Data center Opex is $0.04/W per month.

- The data center has a power usage effectiveness (PUE) of 1.5, the current industry average.

---

[6] We used 12 years in the first edition of the book, but 20 is more consistent with today's industry financial accounting practices.

- Server lifetime is 3 years, and server repair and maintenance is 5% of Capex per year.

- The server's average power draw is 75% of peak power.

Figure 6.1 shows a breakdown of the yearly TCO for case A among data center and server-related Opex and Capex components.

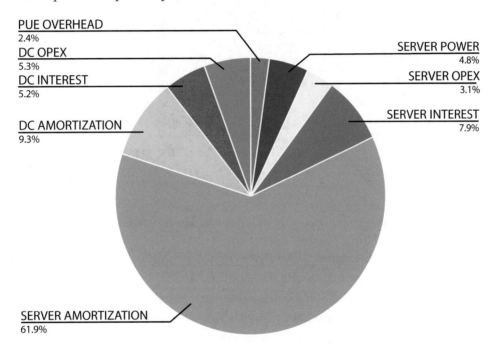

Figure 6.1: TCO cost breakdown for case study A.

In this example, typical of classic data centers, the high server capital costs dominate overall TCO, with 64% of the monthly cost related to server purchase and maintenance. However, commodity-based lower-cost (and perhaps lower-reliability) servers, or higher power prices, can change the picture quite dramatically.

For case B (Figure 6.2), we assume a cheaper, faster, higher-powered server consuming 600 W at peak and costing only $2,000 in a location where the electricity cost is $0.10/kWh. In this case, data center-related costs rise to 44% of the total, and energy costs rise to 19%, with server costs falling to 36%. In other words, the server's hosting cost (that is, the cost of all infrastructure and power to house it) is almost twice the cost of purchasing and maintaining the server in this scenario.

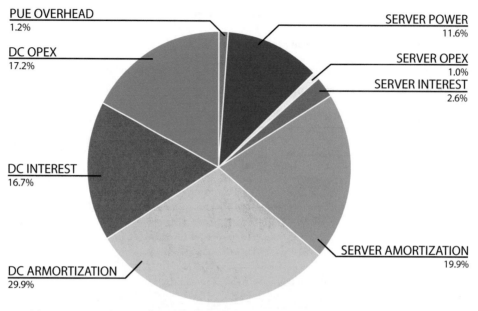

Figure 6.2: TCO cost breakdown for case study B (lower-cost, higher-power servers).

Note that even with the assumed higher power price and higher server power, the absolute 3-year TCO in case B is lower than in case A ($6,310 vs. $7,812) because the server is so much cheaper. The relative importance of power-related costs may increase, as shown in case B, because the power consumption (and performance) of CPUs has more than doubled between 2010 and 2018 (about 14% annually) [Techa], whereas the sale price of low-end servers has stayed relatively stable. As a result, the dollars per watt cost of server hardware is trending down, whereas electricity and construction costs are trending up. This indicates that over the long term, data center facility costs, which are proportional to power consumption, will become a larger fraction of total cost.

6.4   REAL-WORLD DATA CENTER COSTS

In fact, real-world data center costs are even higher than modeled so far. All of the models presented so far assume that the data center is 100% full and the servers are fairly busy (75% of peak power corresponds to a CPU utilization of approximately 50%; see Chapter 5). In reality, this is often not the case. For example, because data center space takes a while to build, we may want to keep a certain amount of empty space to accommodate future deployments. In addition, server layouts assume overly high (worst case) power consumption. For example, a server may consume *up to* 500 W with all options installed (maximum memory, disk, PCI cards, and so on), but the actual configuration deployed may only use 300 W. If the server layout assumes the nameplate rating of 500 W, we will reach a utilization factor of only 60% and thus the actual data center cost per server increases by 1.66x. Thus, in reality, the actual monthly cost per server is often considerably higher than shown above because the data center-related costs increase inversely proportional to data center power utilization.

As discussed in Chapter 5, reaching a high data center power utilization is not as simple as it may seem. Even if the vendor provides a power calculator to compute the actual maximum power draw for a particular configuration, that value will assume 100% CPU utilization. If we install servers based on that value and they run at only 30% CPU utilization on average (consuming 200 W instead of 300 W), we just stranded 30% of the data center capacity. However, if we install servers based on the average value of 200 W and at month's end the servers actually run at near full capacity for a while, our data center will overheat or trip a breaker. Similarly, we may choose to add additional RAM or disks to servers at a later time, which would require physical decompaction of server racks if we left no slack in our power consumption calculations. Thus, in practice, data center operators leave a fair amount of slack space to guard against these problems. Reserves of 20–50% are common, which means that real-world data centers rarely run at anywhere near their rated capacity. In other words, a data center with 10 MW of critical power will often consume a monthly average of just 4–6 MW of actual critical power (plus PUE overhead).

6.5   MODELING A PARTIALLY FILLED DATA CENTER

To model a partially filled data center, we simply scale the Capex and Opex costs (excluding power) by the inverse of the occupancy factor. For example, a data center that is only two-thirds full has a 50% higher Opex. Taking case B above but with a 50% occupancy factor, data center costs completely dominate the cost (Figure 6.3), with only 25% of total cost related to the server. Given the need for slack power just discussed, this case is not as far-fetched as it may sound. Thus, improving actual data center usage (using power capping, for example) can substantially reduce real-world data center costs. In absolute dollars, the server TCO in a completely full data center is $6,310 versus $8,981 in a half-full data center—all that for a server that we assumed cost just $2,000 to purchase!

Partially used servers also affect operational costs in a positive way because the servers use less power. Of course, the savings are questionable because the applications running on those servers are likely to produce less value. Our TCO model cannot capture this effect because it is based on the cost of the physical infrastructure only and excludes the application running on this hardware. To measure this end-to-end performance, we can measure a proxy for application value (for example, the number of bank transactions completed or the number of web searches) and divide the TCO by that number. For example, if we had a data center costing $1 million per month and completing 100 million transactions per month, the cost per transaction would be 1 cent. On the other hand, if traffic is lower at one month and we complete only 50 million transactions, the cost per transaction doubles to 2 cents. In this chapter, we have focused exclusively on hardware costs, but it is important to keep in mind that, ultimately, software performance and server utilization matter just as much. Such issues are also exacerbated in the context of accelerators that deliver higher value but also incur additional costs for the software ecosystem support.

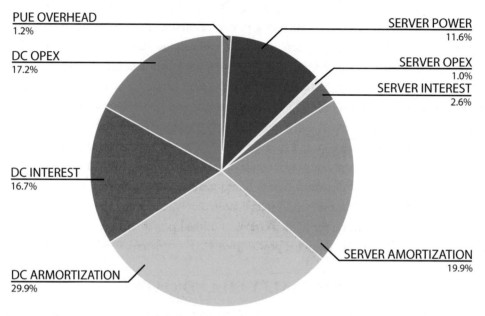

Figure 6.3: TCO case study C (partly filled facility).

6.6   THE COST OF PUBLIC CLOUDS

Instead of building your own data center and server, you can rent a VM from a public cloud provider such as Google's Compute Engine or Amazon's EC2. The Dell server used in our example is roughly comparable to a GCE n1-standard-16 instance, assumed at $0.76/hr as an on-demand instance, or $0.34/hr with a 3-year commitment.

Before we compare these with our cost model, consider the two very different pricing plans. Spot pricing is "pay-as-you-go." You can start and stop a VM at any time, so if you need one for only a few days a year, on-demand pricing will be vastly cheaper than other alternatives. For example, you may need two servers to handle your peak load for six hours per day on weekdays, and one server during the rest of the year. With a spot instance, you pay for only 30 hr per week vs. 168 hr if you owned the server. However, spot instances are fairly expensive: at $0.76/hr, using one for three years at full price will cost $19,972, vs. roughly $8,000 for an owned server. (Note, however, that this does not burden the costs for other factors like utilization as discussed earlier.)

If you need a server for an extended period, public cloud providers will lower the hourly price in exchange for a long-term commitment and an upfront fee. Using the previous three-year contract as an example, a fully utilized instance would cost $8,987, about 45% of what you would pay for an on-demand instance for three years. This is competitive with the cost of an owned machine, possibly even cheaper since you could further reduce your cost in case you didn't need the server after year two.

How can a public cloud provider (who must make a profit on these prices) compete with your in-house costs? In one word: scale. As discussed in this chapter, many operational expenses are relatively independent of the size of the data center: if you want a security guard or a facilities technician on-site 24x7, the cost is the same whether your site is 1 MW or 5 MW. Furthermore, a cloud provider's capital expenses for servers and buildings likely are lower than yours, since they buy (and build) in volume. Google, for example, designs its own servers and data centers to reduce cost.

Why are on-demand instances so much more expensive? Since the cloud provider doesn't know whether you're going to need a server, it keeps additional servers ready in case someone wants them. Thus, the utilization of the server pool used for on-demand instances is substantially below 100% on average. For example, if the typical on-demand instance covers the six hours a day when traffic peaks, their utilization will be 25%, and thus their cost per hour is four times higher than that of a "baseload" instance that runs 24 hr a day.

Many cloud providers offer sustained use discounts to automatically lower the cost of a VM if it is used more continuously throughout a month. For example (GCE), the first 25% of hours in a given month are charged at the full rate, and afterwards the effective rate decreases successively so that a VM used for the entire month receives a 30% overall discount. Such automatic discounts can simplify planning since there is no need to commit to an annual contract ahead of time.

CHAPTER 7

Dealing with Failures and Repairs

The promise of web-based, service-oriented computing will be fully realized only if users can trust that the services they increasingly rely on will be always available. This expectation translates into a high-reliability requirement for building-sized computers. Determining the appropriate level of reliability is fundamentally a tradeoff between the cost of failures (including repairs) and the cost of preventing them. For traditional enterprise-class servers, the cost of failures is thought to be high, and thus designers go to great lengths to provide more reliable hardware by adding redundant power supplies, fans, error correction coding (ECC), RAID disks, and so on. Many legacy enterprise applications were not designed to survive frequent hardware faults, and it is hard to make them fault-tolerant after the fact. Under these circumstances, making the hardware very reliable becomes a justifiable alternative.

In WSCs, however, hardware reliability alone cannot deliver sufficient availability primarily due to its scale. Suppose that a cluster has ultra-reliable server nodes with a stellar mean time between failures (MTBF) of 30 years (10,000 days)—well beyond what is typically possible to achieve at a realistic cost. Even with these ideally reliable servers, a cluster of 10,000 servers will see an average of one server failure per day. Thus, any application that depends on the availability of the entire cluster will see an MTBF no better than one day. In reality, typical servers see an MTBF substantially less than 30 years, and thus the real-life cluster MTBF would be in the range of a few hours between failures. Moreover, large and complex internet services are often composed of several software modules or layers that are not bug-free and can themselves fail at even higher rates than hardware components. Consequently, WSC applications must work around failed servers in software, either with code in the application itself or via functionality provided by middleware, such as a provisioning system for virtual machines that restarts a failed VM on a spare node. Some of the implications of writing software for this environment are discussed by Hamilton [Ham07], based on experience on designing and operating some of the largest services at MSN and Windows Live.

Before we continue, it's important to understand the difference between availability, unavailability, and failure. A system's availability is the fraction of time during which it is available for use; conversely, its unavailability is the fraction of time during which the system isn't available for some reason. Failures are one cause of unavailability, but are often much less common than other causes such as planned maintenance for hardware or software upgrades. Thus, a system with zero failures may still have availability of less than 100%, and a system with a high failure rate may have better availability than one with low failures, if other sources of unavailability dominate.

7.1   IMPLICATIONS OF SOFTWARE FAULT TOLERANCE

Fault-tolerant software is inherently more complex than software that can assume fault-free operation. As much as possible, we should try to implement a fault-tolerant software infrastructure layer that can hide much of the complexity of dealing with failures from application-level software.

There are some positive consequences of adopting a fault-tolerant model, though. Once hardware faults can be tolerated without undue disruption to a service, computer architects have some leeway to choose the level of hardware reliability that maximizes overall system cost efficiency. This leeway enables consideration, for instance, of using inexpensive PC-class hardware for a server platform instead of mainframe-class computers, as discussed in Chapter 3. In addition, this model can lead to simplifications in common operational procedures. For example, to upgrade system software in a cluster, you can load a newer version in the background (that is, during normal operation), kill the older version, and immediately start the newer one. Hardware upgrades can follow a similar procedure. Basically, the same fault-tolerant software infrastructure mechanisms built to handle server failures could have all the required mechanisms to support a broad class of operational procedures. By choosing opportune time windows and rate-limiting the pace of kill–restart actions, operators can still manage the desired number of planned service-level disruptions.

The basic property being exploited here is that, unlike in traditional server setups, it is no longer necessary to keep a server running at all costs. This simple requirement shift affects almost every aspect of the deployment, from machine and data center design to operations, often enabling optimization opportunities that would not be on the table otherwise. For instance, let us examine how this affects the recovery model. A system that needs to be highly reliable in the presence of unavoidable transient hardware faults, such as uncorrectable errors caused by cosmic particle strikes, may require hardware support for checkpoint recovery so that upon detection the execution can be restarted from an earlier correct state. A system that is allowed to go down upon occurrence of such faults may choose not to incur the extra overhead in cost or energy of checkpointing.

Another useful example involves the design tradeoffs for a reliable storage system. One alternative is to build highly reliable storage nodes through the use of multiple disk drives in a mirrored or RAIDed configuration so that a number of disk errors can be corrected on the fly. Drive redundancy increases reliability but by itself does not guarantee that the storage server will be always up. Many other single points of failure also need to be attacked (such as power supplies and operating system software), and dealing with all of them incurs extra cost while never assuring fault-free operation. Alternatively, data can be mirrored or RAIDed across disk drives that reside in multiple machines—the approach chosen by Google's GFS or Colossus file systems [GGL03]. This option tolerates not only drive failures but also entire storage server crashes because other replicas of each piece of data are accessible through other servers. It also has different performance characteristics from the centralized storage server scenario. Data updates may incur higher networking overheads because they require communicating with multiple systems to update all replicas, but aggregate read bandwidth can be greatly increased because clients can source data from multiple endpoints (in the case of full replication).

In a system that can tolerate a number of failures at the software level, the minimum requirement made to the hardware layer is that its faults are always detected and reported to software in a timely enough manner as to allow the software infrastructure to contain it and take appropriate recovery actions. It is not necessarily required that hardware transparently correct all faults. This does not mean that hardware for such systems should be designed without error correction capabilities. Whenever error correction functionality can be offered *within a reasonable cost or complexity*, it often pays to support it. It means that if hardware error correction would be exceedingly expensive, the system would have the option of using a less expensive version that provided *detection* capabilities only. Modern DRAM systems are a good example of a case in which powerful error *correction* can be provided at a low additional cost.

Relaxing the requirement that hardware errors be detected, however, would be much more difficult because every software component would be burdened with the need to check its own correct execution. At one early point in its history, Google had to deal with servers whose DRAM lacked even parity checking. Producing a web search index consists essentially of a very large shuffle/merge sort operation, using several machines over a long period. In 2000, one of the then monthly updates to Google's web index failed pre-release checks when a subset of tested queries was found to return seemingly random documents. After some investigation a pattern was found in the new index files that corresponded to a bit being stuck at zero at a consistent place in the data structures; a bad side effect of streaming a lot of data through a faulty DRAM chip. Consistency checks were added to the index data structures to minimize the likelihood of this problem recurring, and no further problems of this nature were reported. Note, however, that this workaround did not guarantee 100% error detection in the indexing pass because not all memory positions were being checked—instructions, for example, were not. It worked because index data structures were so much larger than all other data involved in the computation that having those self-checking data structures made it very likely that machines with defective DRAM would be identified and excluded from the cluster. The next machine generation at Google did include memory parity detection, and once the price of memory with ECC dropped to competitive levels, all subsequent generations have used ECC DRAM.

7.2   CATEGORIZING FAULTS

An efficient fault-tolerant software layer must be based on a set of expectations regarding fault sources, their statistical characteristics, and the corresponding recovery behavior. Software developed in the absence of such expectations can suffer from two risks: being prone to outages if the underlying faults are underestimated, or requiring excessive overprovisioning if faults are assumed to be much more frequent than they actually are.

Providing an accurate quantitative assessment of faults in WSC systems is challenging given the diversity of equipment and software infrastructure across different deployments. Instead, we will attempt to summarize the high-level trends from publicly available sources and from our own experience.

7.2.1 FAULT SEVERITY

Hardware or software faults can affect internet services in varying degrees, resulting in different service-level failure modes. The most severe modes may demand high reliability levels, whereas the least damaging modes might have more relaxed requirements that can be achieved with less expensive solutions. We broadly classify service-level failures into the following categories, listed in decreasing degree of severity.

- *Corrupted*: Committed data are impossible to regenerate, lost, or corrupted.

- *Unreachable*: Service is down or otherwise unreachable by users.

- *Degraded*: Service is available but in some degraded mode.

- *Masked*: Faults occur but are completely hidden from users by fault-tolerant software and hardware mechanisms.

Acceptable levels of robustness will differ across those categories. We expect most faults to be masked by a well-designed fault-tolerant infrastructure so that they are effectively invisible outside of the service provider. It is possible that masked faults will impact the service's maximum sustainable throughput capacity, but a careful degree of overprovisioning can ensure that the service remains healthy.

If faults cannot be completely masked, their least severe manifestation is one in which there is some degradation in the quality of service. Here, different services can introduce degraded availability in different ways. One example of such degraded service is when a web search system uses data partitioning techniques to improve throughput but loses some systems that serve parts of the database [Bre01]. Search query results will be imperfect but probably still acceptable in many cases. Graceful degradation as a result of faults can also manifest itself as decreased freshness. For example, a user may access his or her email account, but new email delivery is delayed by a few minutes, or some fragments of the mailbox could be temporarily missing. Although these kinds of faults also need to be minimized, they are less severe than complete unavailability. Internet services need to be deliberately designed to take advantage of such opportunities for gracefully degraded service. In other words, this support is often application-specific and not easily hidden within layers of cluster infrastructure software.

Service availability and reachability are very important, especially because internet service revenue is often related in some way to traffic volume [Cha+01b]. However, perfect availability is not a realistic goal for internet-connected services because the internet itself has limited availability characteristics. Chandra et al. [Cha+01b] report that internet endpoints may be unable to reach each other between 1% and 2% of the time due to a variety of connectivity problems, including routing issues. That translates to an availability of less than "two nines." In other words, even if an internet service is perfectly reliable, users will, on average, perceive it as being no greater than 99.0% available. As a result, an internet-connected service that avoids long-lasting outages for any large group of users and has an average unavailability of less than 1% will be difficult to distinguish from a perfectly reliable system. Google measurements of internet availability as of 2014 indicated that it was likely on average in the range of 99.6–99.9% when Google servers are one of the endpoints, but the spectrum is fairly wide. Some areas of the world experience significantly lower availability.

Measuring service availability in absolute time is less useful for internet services that typically see large daily, weekly, and seasonal traffic variations. A more appropriate availability metric is the fraction of requests satisfied by the service divided by the total number of requests made by users; a metric called *yield* by Brewer [Bre01].

Finally, one particularly damaging class of failures is the loss or corruption of committed updates to critical data, particularly user data, critical operational logs, or relevant data that are hard or impossible to regenerate. Arguably, it is much more critical for services not to lose data than to be perfectly available to all users. It can also be argued that such critical data may correspond to a relatively small fraction of all the data involved in a given service operation. For example, copies of the web and their corresponding index files are voluminous and important data for a search engine, but can ultimately be regenerated by recrawling the lost partition and recomputing the index files.

In summary, near perfect reliability is not universally required in internet services. Although it is desirable to achieve it for faults such as critical data corruption, most other failure modes can tolerate lower reliability characteristics. Because the internet itself has imperfect availability, a user may be unable to perceive the differences in quality of service between a perfectly available service and one with, say, four 9s (99.99%) of availability.

7.2.2 CAUSES OF SERVICE-LEVEL FAULTS

In WSCs, it is useful to understand faults in terms of their likelihood of affecting the health of the whole system, such as causing outages or other serious service-level disruption. Oppenheimer et al. [OGP03] studied three internet services, each consisting of more than 500 servers, and tried to identify the most common sources of service-level failures. They conclude that operator-caused or misconfiguration errors are the largest contributors to service-level failures, with hardware-related faults (server or networking) contributing to 10–25% of the total failure events.

Oppenheimer's results are somewhat consistent with the seminal work by Gray [Gra90], which doesn't look at internet services but instead examines field data from the highly fault-tolerant Tandem servers between 1985 and 1990. He also finds that hardware faults are responsible for a small fraction of total outages (less than 10%). Software faults (~60%) and maintenance/operations faults (~20%) dominate the outage statistics.

It is surprising at first to see hardware faults contributing to so few outage events in these two widely different systems. Rather than making a statement about the underlying reliability of the hardware components in these systems, such numbers indicate how successful the fault-tolerant techniques have been in preventing component failures from affecting high-level system behavior. In Tandem's case, such techniques were largely implemented in hardware, whereas in the systems Oppenheimer studied, we can attribute it to the quality of the fault-tolerant software infrastructure. Whether software- or hardware-based, fault-tolerant techniques do particularly well when faults are largely statistically independent, which is often (even if not always) the case in hardware faults. Arguably, one important reason why software-, operator-, and maintenance-induced faults have a high impact on outages is because they are more likely to affect multiple systems at once, thus creating a correlated failure scenario that is much more difficult to overcome.

Our experience at Google is generally in line with Oppenheimer's classification, even if the category definitions are not fully consistent. Figure 7.1 represents a rough classification of all events that corresponded to noticeable disruptions at the service level in one of Google's large-scale online services. These are not necessarily outages (in fact, most of them are not even user-visible events), but correspond to situations where some kind of service degradation is noticed by the monitoring infrastructure and must be scrutinized by the operations team. As expected, the service is less likely to be disrupted by machines or networking faults than by software errors, faulty configuration data, and human mistakes.

Factors other than hardware equipment failure dominate service-level disruption because it is easier to architect services to tolerate known hardware failure patterns than to be resilient to general software bugs or operator mistakes. A study by Jiang et al. [Jia+08], based on data from over 39,000 storage systems, concludes that disk failures are in fact not a dominant contributor to storage system failures. That result is consistent with analysis by Ford et al. [For+10] of distributed storage systems availability. In that study, conducted using data from Google's Colossus distributed file system, planned storage node reboot events are the dominant source of node-level unavailability. That study also highlights the importance of understanding correlated failures (failures in multiple storage nodes within short time windows), as models that don't account for correlation can underestimate the impact of node failures by many orders of magnitude.

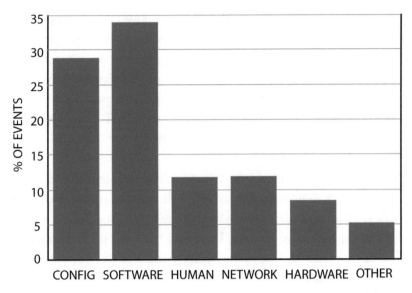

Figure 7.1: Distribution of service disruption events by most likely cause at one of Google's main services, collected over a period of six weeks by Google's Robert Stroud.

7.3   MACHINE-LEVEL FAILURES

An important factor in designing fault-tolerant distributed systems is understanding availability at the server level. Here we consider machine-level failures to be any situation that leads to a server being down, whatever the cause (such as operating system bugs).

As with cluster-service failures, relatively little published field data exists on server availability. A 1999 study by Kalyanakrishnam et al. [KKI99] finds that Windows NT machines involved in a mail routing service for a commercial organization were on average 99% available. The authors observed 1,100 reboot events across 66 servers and saw an average uptime of 11.82 days (median of 5.54 days) and an average downtime of just less than 2 hr (median of 11.43 min). About half of the reboots were classified as abnormal; that is, were due to a system problem instead of a normal shutdown. Only 10% of the reboots could be blamed on faulty hardware or firmware. The data suggest that application faults, connectivity problems, or other system software failures are the largest known crash culprits. If we are interested only in the reboot events classified as abnormal, we arrive at an MTTF of approximately 22 days, or an annualized machine failure rate of more than 1,600%.

Schroeder and Gibson [SG07a] studied failure statistics from high-performance computing systems at Los Alamos National Laboratory. Although these are not a class of computers that we are interested in here, they are made up of nodes that resemble individual servers in WSCs, so their data is relevant in understanding machine-level failures in our context. Their analysis spans nearly 24,000 processors, with more than 60% of them deployed in clusters of small-scale SMPs (2–4 processors per node). Although the node failure rates vary by more than a factor of 10x across different systems, the failure rate normalized by number of processors is much more stable—approximately 0.3 faults per year per CPU—suggesting a linear relationship between the number of sockets and unreliability. If we assume servers with four CPUs, we could expect machine-level failures to be at a rate of approximately 1.2 faults per year or an MTTF of approximately 10 months. This rate of server failures is more than 14 times lower than the one observed in Kalyanakrishnan's study [KKI99].

Google's machine-level failure and downtime statistics are summarized in Figures 7.2 and 7.3. The data is based on a six-month observation of all machine restart events and their corresponding downtime, where downtime corresponds to the entire time interval where a machine is not available for service, regardless of cause. These statistics cover all of Google's machines. For example, they include machines that are in the repairs pipeline, planned downtime for upgrades, as well as all kinds of machine crashes.

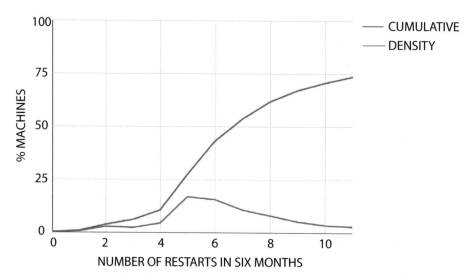

Figure 7.2: Distributions of machine restarts over six months at Google. (Updated in 2018.)

Figure 7.2 shows the distribution of machine restart events. The graph shows that 50% or more machines restart at least once a month, on average. The tail is relatively long (the figure truncates the data at 11 or more restarts) due to the large population of machines in Google's fleet. Approximately 5% of all machines restart more than once a week. Several effects, however, are smudged away by such large-scale averaging. For example, we typically see higher than normal failure rates during the first few months of new server product introduction. The causes include manufacturing bootstrapping effects, firmware and kernel bugs, and occasional hardware problems that become noticeable only after a large number of systems are in use. If we exclude from the sample all machines that are still suffering from such effects, the annualized restart rate corresponds to approximately one month between restarts, on average, largely driven by Google's bug resolution, feature enhancements, and security-processes-related upgrades. We also note that machines with frequent restarts are less likely to be in active service for long.

In another internal study conducted recently, planned server unavailability events were factored out from the total machine unavailability. The remaining unavailability, mainly from server crashes or lack of networking reachability, indicates that a server can stay up for an average of nearly two years (0.5 unplanned restart rate). This is consistent with our intuition that most restarts are due to planned events, such as software and hardware upgrades.

These upgrades are necessary to keep up with the velocity of kernel changes and also allows Google to prudently react to emergent & urgent security issues. As discussed earlier, it is also important to note that Google Cloud's Compute Engine offers live migration to keep the VM instances running by migrating the running instances to another host in the same zone rather than requiring your VMs to be rebooted. Note that live migration does not change any attributes or properties of the VM itself.

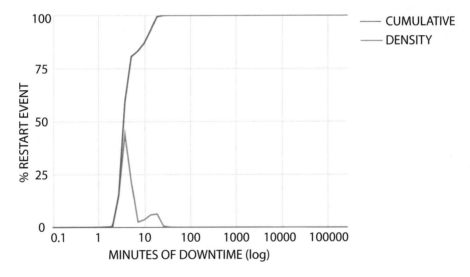

Figure 7.3: Distribution of machine downtime, observed at Google over six months. The average annualized restart rate across all machines is 12.4, corresponding to a mean time between restarts of just less than one month.

Restart statistics are key parameters in the design of fault-tolerant software systems, but the availability picture is complete only once we combine it with downtime data—a point articulated earlier by the Berkeley ROC project [Pat+02]. Figure 7.3 shows the distribution of downtime from the same population of Google servers. The x-axis displays downtime, against both density and cumulative machine distributions. Note that the data include both planned reboots and those caused by miscellaneous hardware and software failures. Downtime includes all the time a machine stopped operating until it has rebooted and basic node services were restarted. In other words, the downtime interval ends not when the machine finishes rebooting, but when key basic daemons are up.

Approximately 55% of all restart events last less than 3 min, 25% of them last between 3 and 30 min, with most of the remaining restarts finishing in about a day. These usually correspond to some combination of physical repairs (SW directed swaps) and automated file system recovery from crashes. Approximately 1% of all restart events last more than a day, usually corresponding to systems going into repairs. The average downtime is just more than 10 min. The resulting average machine availability is 99.93%.

When provisioning fault-tolerant software systems, it is also important to focus on real (unexpected) machine crashes, as opposed to the previous analysis that considers all restarts. In our experience, the crash rate of mature servers (those that survived infant mortality) ranges between 1.2 and 2 crashes per year. In practice, this means that a service that uses 2,000 servers should plan to operate normally while tolerating a machine crash approximately every 2.5 hr, or approximately 10 machines per day. Given the expected machine downtime for 99% of all restart cases is less than 2 days, one would need as few as 20 spare machines to safely keep the service fully provisioned. A larger margin might be desirable if there is a large amount of state that must be loaded before a machine is ready for service.

7.3.1   WHAT CAUSES MACHINE CRASHES?

Reliably identifying culprits for machine crashes is generally difficult because, in many situations, transient hardware errors can be hard to distinguish from operating system or firmware bugs. However, there is significant indirect and anecdotal evidence suggesting that software-induced crashes are much more common than those triggered by hardware faults. Some of this evidence comes from component-level diagnostics. Because memory and disk subsystem faults were the two most common diagnostics for servers sent to hardware repairs within Google in 2018, we will focus on those.

DRAM Soft Errors

Although there are little available field data on this topic, it is generally believed that DRAM soft error rates are extremely low once modern ECCs are used. In a 1997 IBM white paper, Dell [Del97] sees error rates from chipkill ECC being as low as six errors for 10,000 one-GB systems over three years (0.0002 errors per GB per year—an extremely low rate). A survey article by Tezzaron Semiconductor in 2004 [Terra] concludes that single-error rates per Mbit in modern memory devices range between 1,000 and 5,000 FITs (failures in time, defined as the rate of faults per billion operating hours), but that the use of ECC can drop soft-error rates to a level comparable to that of hard errors.

A study by Schroeder et al. [SPW09] evaluated DRAM errors for the population of servers at Google and found FIT rates substantially higher than previously reported (between 25,000 and 75,000) across multiple DIMM technologies. That translates into correctable memory errors affecting about a third of Google machines per year and an average of one correctable error per server every 2.5 hr. Because of ECC technology, however, only about 1.3% of all machines ever experience uncorrectable memory errors per year. A more recent study [HSS12] found that a large fraction of DRAM errors could be attributed to hard (non-transient) errors and suggested that simple page retirement policies could mask a large fraction of DRAM errors in production systems while sacrificing only a negligible fraction of the total DRAM in the system.

Disk Errors

Studies based on data from NetApp and the University of Wisconsin [Bai+07], Carnegie Mellon [SG07b], and Google [PWB07] have recently examined the failure characteristics of modern disk drives. Hard failure rates for disk drives (measured as the annualized rate of replaced components) have typically ranged between 2% and 4% in large field studies, a much larger number than the usual manufacturer specification of 1% or less. Bairavasundaram et al. [Bai+07] looked specifically at the rate of latent sector errors—a measure of data corruption frequency. In a population of more than 1.5 million drives, they observed that less than 3.5% of all drives develop any errors over a period of 32 months.

These numbers suggest that the average fraction of machines crashing annually due to disk or memory subsystem faults should be less than 10% of all machines. Instead, we observe crashes to be more frequent and more widely distributed across the machine population. We also see noticeable variations on crash rates within homogeneous machine populations that are more likely explained by firmware and kernel differences.

The effect of ambient temperature on the reliability of disk drives has been well studied by Pinheiro et al. [PWB07] and El Sayed et al. [ES+]. While common wisdom previously held that temperature had an exponentially negative effect on the failure rates of disk drives, both of these field studies found little or no evidence of that in practice. In fact, both studies suggest that most disk errors appear to be uncorrelated with temperature.

Another indirect evidence of the prevalence of software-induced crashes is the relatively high mean time to hardware repair observed in Google's fleet (more than six years) when compared to the mean time to machine crash (six months or less).

It is important to mention that a key feature of well-designed fault-tolerant software is its ability to survive individual faults, whether they are caused by hardware or software errors. One architectural option that can improve system reliability in the face of disk drive errors is the trend toward diskless servers; once disks are a networked resource it is easier for a server to continue operating by failing over to other disk devices in a WSC.

7.3.2 PREDICTING FAULTS

The ability to predict future machine or component failures is highly valued because it could avoid the potential disruptions of unplanned outages. Clearly, models that can predict most instances of a given class of faults with very low false-positive rates can be very useful, especially when those predictions involve short time-horizons—predicting that a memory module will fail within the next 10 years with 100% accuracy is not particularly useful from an operational standpoint.

When prediction accuracies are less than perfect, which unfortunately tends to be true in most cases, the model's success will depend on the tradeoff between accuracy (both in false-positive rates and time horizon) and the penalties involved in allowing faults to happen and recovering from them. Note that a false component failure prediction incurs all of the overhead of the regular hardware repair process (parts, technician time, machine downtime, etc.). Because software in WSCs is designed to gracefully handle all the most common failure scenarios, the penalties of letting faults happen are relatively low; therefore, prediction models must have much greater accuracy to be economically competitive. By contrast, traditional computer systems in which a machine crash can be very disruptive to the operation may benefit from less accurate prediction models.

Pinheiro et al. [PWB07] describe one of Google's attempts to create predictive models for disk drive failures based on disk health parameters available through the Self-Monitoring Analysis and Reporting Technology (SMART) standard. They conclude that such models are unlikely to predict individual drive failures with sufficient accuracy, but they can be useful in reasoning about the expected lifetime of groups of devices which can be useful in optimizing the provisioning of replacement units.

7.4 REPAIRS

An efficient repair process is critical to the overall cost efficiency of WSCs. A machine in repair is effectively out of operation, so the longer a machine is in repair, the lower the overall availability of the fleet. Also, repair actions are costly in terms of both replacement parts and the skilled labor involved. Last, repair quality—how likely a repair action will actually fix a problem while accurately determining which (if any) component is at fault—affects both component expenses and average machine reliability.

Two characteristics of WSCs directly affect repair efficiency. First, because of the high number of relatively low-end servers involved and the presence of a software fault-tolerance layer, quickly responding to individual repair cases is not as critical because the repairs are unlikely to affect overall service health. Instead, a data center can implement a schedule that makes the most efficient use of a technician's time by making a daily sweep of all machines that need repairs attention. The philosophy is to increase the rate of repairs while keeping the repair latency within acceptable levels.

In addition, when many thousands of machines are in operation, massive volumes of data about machine health can be collected and analyzed to create automated systems for health determination and diagnosis. Google's system health infrastructure, illustrated in Figure 7.4, is an example of a monitoring system that takes advantage of this massive data source. It constantly monitors every server for configuration, activity, environmental, and error data. This information is stored as a time series in a scalable repository where it can be used for various kinds of analysis, including an automated machine failure diagnostics tool that uses machine learning methods to suggest the most appropriate repairs action.

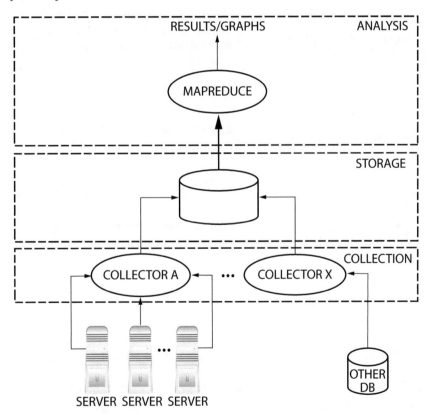

Figure 7.4: Google's system health monitoring and analysis infrastructure.

In addition to performing individual machine diagnostics, the system health infrastructure is useful in other ways. For example, it monitors the stability of new system software versions and has helped pinpoint a specific batch of defective components across a large fleet of machines. It has also been valuable in longer-term analytical studies, such as the disk failure study by Pinheiro et al., mentioned in the previous section, and the data center-scale power provisioning study by Fan et al. [FWB07].

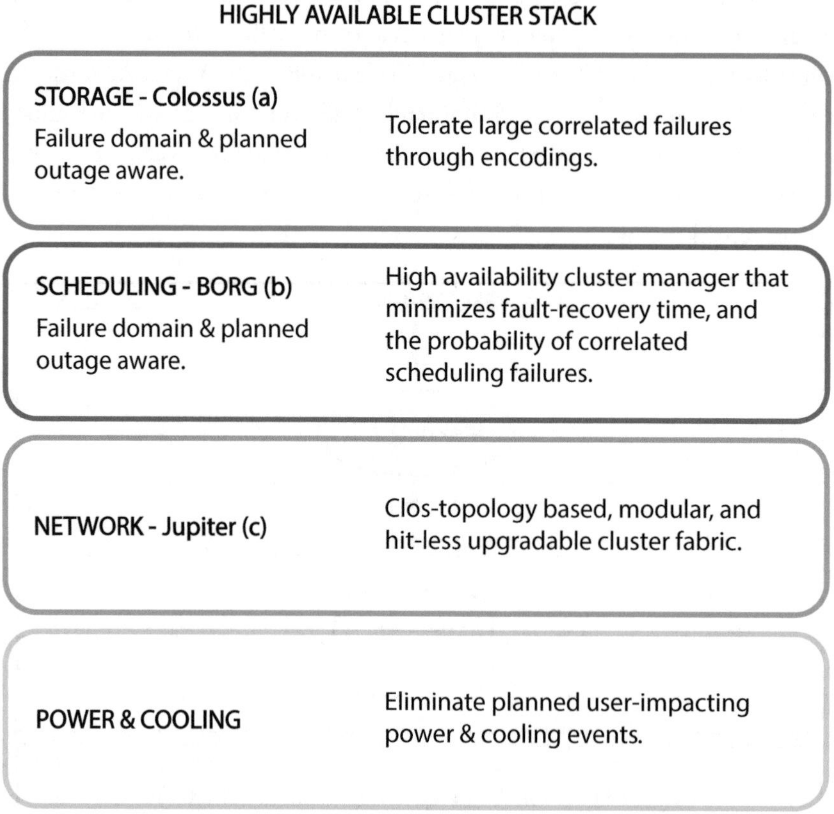

HIGHLY AVAILABLE CLUSTER STACK

STORAGE - Colossus (a)	Tolerate large correlated failures through encodings.
Failure domain & planned outage aware.	

SCHEDULING - BORG (b)	High availability cluster manager that minimizes fault-recovery time, and the probability of correlated scheduling failures.
Failure domain & planned outage aware.	

NETWORK - Jupiter (c)	Clos-topology based, modular, and hit-less upgradable cluster fabric.

POWER & COOLING	Eliminate planned user-impacting power & cooling events.

Figure 7.5: Highly available cluster architecture [Ser17, Ver+15, Sin+15].

7.5 TOLERATING FAULTS, NOT HIDING THEM

The capacity of well-designed fault-tolerant software to mask large numbers of failures with relatively little impact to service-level metrics could have unexpectedly dangerous side effects. Consider a three-tier application representing a web service with the backend tier replicated three times. Such replicated setups have the dual purpose of increasing peak throughput as well as tolerating server faults when operating below peak capacity. Assume that the incoming request rate is at 50% of total capacity. At this level, this setup could survive one backend failure with little disruption in service levels. However, a second backend failure would have a dramatic service-level impact that could theoretically result in a complete outage.

In systems descriptions, we often use N to denote the number of servers required to provide a service at full load, so N + 1 describes a system with one additional replica for redundancy. Common arrangements include N (no redundancy), N + 1 (tolerating a single failure), N + 2 ("concurrently maintainable," tolerating a single failure even when one unit is offline for planned maintenance), and 2N (mirroring of every unit).

Systems with large amounts of internal replication to increase capacity (horizontal scaling) provide redundancy at a very low cost. For example, if we need 100 replicas to handle the daily peak load, an N + 2 setup incurs just 2% overhead for high availability. Such systems can tolerate failures so well that an outside observer might be unaware of how much internal slack remains, or in other words, how close to the edge one might be. In those cases, the transition from healthy behavior to meltdown can be abrupt, which is not a desirable property. This example emphasizes the importance of comprehensive monitoring, both at the application (or service) level as well as the machine infrastructure level, so that faults can be well tolerated and yet visible to operators. This enables prompt corrective action when the amount of internal redundancy approaches the limits of what the fault-tolerant software layer can handle.

Still, broken machines eventually must be repaired. Traditional scenarios, where a repair must happen immediately, require more costly staging of replacement parts as well as additional costs to bringing a service technician on site. When we can batch repairs, we can lower these costs per repair. For reference, service contracts for IT equipment that provide on-site repair within 24 hr typically come at an annual cost of 5–15% of the equipment's value; a 4-hr response time usually doubles that cost. In comparison, repairs in large server farms are cheaper. To illustrate the point, assume that a WSC has enough scale to keep a full-time repairs technician busy. Assuming 1 hr per repair and an annual failure rate of 5%, a system with 40,000 servers would suffice; in reality, that number will be considerably smaller because the same technician can also handle installations and upgrades. Let us further assume that the hourly cost of a technician is $100 and that the average repair requires replacement parts costing 10% of the system cost; both of these assumptions are generously high. Still, for a cluster of servers costing $2,000 each, we arrive at an annual cost per server of 5% * ($100 + 10% * $2,000) = $15, or 0.75% per year. In other words, keeping large clusters healthy can be quite affordable.

7.6   ACCOUNTING FOR FAULTS IN CLUSTER SYSTEM DESIGN

For services with mutable state, as previously noted, replicas in different clusters must now worry about consistent replicas of highly mutable data. However, most new services start serving with a small amount of traffic and are managed by small engineering teams. In such cases these services need only a small fraction of the capacity of a given cluster, but if successful, will grow over time. For these services, even if they do not have highly mutable state, managing multiple clusters imposes a significant overhead, both in terms of engineering time and resource cost; that is, N + 2 replication is expensive when N = 1. Furthermore, these services tend to be crowded out of clusters by the large services and by the fact that a large number of services leads to resource fragmentation. In all likelihood, the free resources are not where the service is currently running.

To address these problems, the system cluster design needs to include a unit of resource allocation and job schedulability with certain availability, and eliminate planned outages as a source of overhead for all but singly-homed services with the highest availability requirements.

Figure 7.5 shows a schematic of such a cluster stack.

Power: At the base of the stack, we tackle the delivery of power and cooling to machines. Typically, we target a fault tolerant and concurrently maintainable physical architecture for the power and cooling system. For example, as described in Chapter 4, power is distributed hierarchically at the granularity of the building and physical data center rows. For high availability, cluster scheduling purposely spreads jobs across the units of failure. Similarly, the required redundancy in storage systems is in part determined by the fraction of a cluster that may simultaneously fail as a result of a power event. Hence, larger clusters lead to lower storage overhead and more efficient job scheduling while meeting diversity requirements.

Networking: A typical cluster fabric architecture, like the Jupiter design described in Chapter 3, achieves high availability by redundancy in fabric and physical diversity in deployment. It also focuses on robust software for the necessary protocols and a reliable out-of-band control plane. Given the spreading requirement for the cluster scheduler as previously discussed, Jupiter also supports uniform bandwidth across the cluster and resiliency mechanisms in the network control plane in addition to data path diversity.

Scheduling: Applications that run on Borg (Google's large-scale cluster management system) are expected to handle scheduling-related failure events using techniques such as replication, storing persistent state in a distributed file system, and (if appropriate) taking occasional checkpoints. Even so, we try to mitigate the impact of these events. For example, Borg scheduling accounts for automatic rescheduling of evicted applications, on a new machine if necessary, and reduced correlated failures by spreading applications across failure domains such as machines, racks, and power domains.

Storage: We keep the overall storage system highly available with two simple yet effective strategies: fast recovery and replication/encoding. Details are listed in the reference [Ser17].

Together, these capabilities provide a highly available cluster suitable for a large number of big or small services. In cloud platforms such as Google Cloud Platform (GCP), high availability is exposed to customers using the concept of zones and regions [GCRZ]. A representative mapping of compute zones to regions is shown in Figure 7.6.

Google internally maintains a map of clusters to zones such that the appropriate product SLAs (Service Level Agreements) can be satisfied (for example, Google Compute Engine SLAs [GCE]). Cloud customers are encouraged to utilize the zone and region abstractions in developing highly available and fault-tolerant applications on GCP.

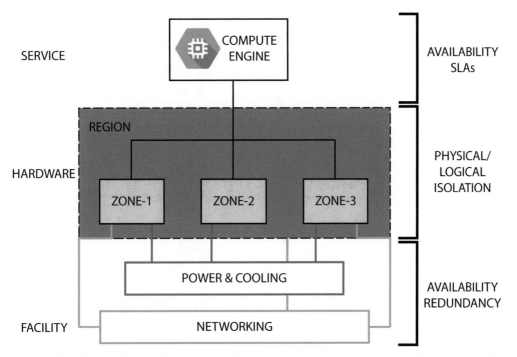

Figure 7.6: Mapping of Google Compute Engine zones to regions, with associated physical and logical isolation and customer facing availability properties.

CHAPTER 8

Closing Remarks

Pervasive Internet access, the proliferation of mobile devices, and the rapid growth of cloud computing have all enabled an ever growing number of applications to move to a web services delivery model ("to the cloud"). In this model, the massive amounts of well-connected processing and storage resources in large datacenters can be efficiently amortized across a large user population and multiple ubiquitous workloads. These datacenters are quite different from traditional colocation or hosting facilities of earlier times, constituting a new class of large-scale computers. The software in these computers is built from several individual programs that interact together to implement complex Internet services, and may be designed and maintained by different teams of engineers, perhaps even across organizational and company boundaries. The data volume manipulated by such computers can range from hundreds to thousands of terabytes, with service-level requirements for high availability, high throughput, and low latency often requiring replication of the baseline data set. Applications of this scale do not run on a single server or even on a rack of servers. They require clusters of many hundreds or thousands of individual servers, with their corresponding storage and networking subsystems, power distribution and conditioning equipment, and cooling infrastructure.

Our central point is simple: this computing platform cannot be viewed simply as a miscellaneous collection of co-located machines. Large portions of the hardware and software resources in these datacenters must work in concert to deliver good levels of Internet service performance, something that can only be achieved by a holistic approach to their design and deployment. In other words, we must treat the datacenter itself as one massive computer. The enclosure for this computer bears little resemblance to a pizza box or a refrigerator, the images chosen to describe servers in past decades. Instead it looks more like a building or warehouse—computer architecture meets traditional (building) architecture—a *warehouse-scale computer* (*WSC*).

Hardware and software architects need to understand the characteristics of this class of computing systems so that they can continue to design and program today's WSCs. WSCs are built from a relatively homogeneous collection of components (servers, storage, and networks) and use a common software management and scheduling infrastructure across all computing nodes to orchestrate resource usage among multiple workloads. In the remainder of this chapter, we summarize the main characteristics of WSC systems described in previous sections and list some important challenges and trends.

8.1 HARDWARE

The building blocks of choice for WSCs are commodity server-class machines, consumer- or enterprise-grade storage devices, and Ethernet-based networking fabrics. Driven by the purchasing volume of hundreds of millions of consumers and small businesses, commodity components benefit from manufacturing economies of scale and therefore present significantly better price/performance ratios than their corresponding high-end counterparts. In addition, Internet and cloud applications tend to exhibit large amounts of easily exploitable parallelism, making the peak performance of an individual server less important than the aggregate throughput of a collection of servers.

The higher reliability of high-end equipment is less important in this domain because a fault-tolerant software layer is required to provision a dependable Internet service regardless of hardware quality. Even if we used highly reliable servers, clusters with tens of thousands of systems will experience failures too frequently for software to assume fault-free operation. Moreover, large and complex Internet services are often composed of multiple software modules or layers that are not bug-free and can fail at even higher rates than hardware components.

Given the baseline reliability of WSC components and the large number of servers used by a typical workload, there are likely no useful intervals of fault-free operation: we must assume that the system is operating in a state of near-continuous recovery. This state is especially challenging for online services that need to remain available every minute of every day. For example, it is impossible to use the recovery model common to many HPC clusters, which pause an entire cluster workload upon an individual node failure and restart the whole computation from an earlier checkpoint. Consequently, WSC applications must work around failed servers in software, either at the application level or (preferably) via functionality provided via middleware, such as a provisioning system for virtual machines that restarts a failed VM on spare nodes. Despite the attractiveness of low-end, moderately reliable server building blocks for WSCs, high-performance, high-availability components still have value in this class of systems. For example, fractions of a workload (such as SQL databases) may benefit from higher-end SMP servers with their larger interconnect bandwidth. However, highly parallel workloads and fault-tolerant software infrastructures effectively broaden the space of building blocks available to WSC designers, allowing lower end options to work very well for many applications.

The performance of the networking fabric and the storage subsystem can be more relevant to WSC programmers than CPU and DRAM subsystems, unlike what is more typical in smaller scale systems. The relatively high costs (per gigabyte) of DRAM or FLASH storage make them prohibitively expensive for large data sets or infrequently accessed data; therefore, disk drives are still used heavily. The increasing gap in performance between DRAM and disks, and the growing imbalance between throughput and capacity of modern disk drives makes the storage subsystem a common performance bottleneck in large-scale systems, motivating broader use of Flash, and potentially new non-volatile memory technologies like ZNAND [Sam17] or 3D Xpoint [3DX]. The use of many small-scale servers demands networking fabrics with very high port counts and high bisection bandwidth. Because such fabrics are costly today, programmers must be keenly aware of the scarcity of datacenter-level bandwidth when architecting software systems. This results in more complex software solutions, expanded design cycles, and sometimes inefficient use of global resources.

8.2   SOFTWARE

WSCs are more complex programming targets than traditional computing systems because of their immense scale, complexity of their architecture (as seen by the programmer), and the need to tolerate frequent failures.

Internet services must achieve high availability, typically aiming for a target of 99.99% or better (about an hour of downtime per year). As mentioned earlier, achieving fault-free operation on a large collection of hardware and system software is infeasible, therefore warehouse-scale workloads must be designed to gracefully tolerate high numbers of component failures with little or no impact on service-level performance and availability.

This workload differs substantially from that running in traditional HPC datacenters, the traditional users of large-scale cluster computing. Like HPC applications, these workloads require significant CPU resources, but the individual tasks are less synchronized than in typical HPC applications and communicate less intensely. Furthermore, they are much more diverse, unlike HPC applications that exclusively run a single binary on a large number of nodes. Much of the parallelism inherent in this workload is natural and easy to exploit, stemming from the many users concurrently accessing the service or from the parallelism inherent in data mining. Utilization varies, often with a diurnal cycle, and rarely reaches 90% because operators prefer to keep reserve capacity for unexpected load spikes (flash crowds) or to take on the load of a failed cluster elsewhere in the world. In comparison, an HPC application may run at full CPU utilization for days or weeks.

Software development for Internet services also differs from the traditional client/server model in a number of ways. First, typical Internet services exhibit ample parallelism stemming from both data parallelism and request-level parallelism. Typically, the problem is not to find parallelism but to manage and efficiently harness the explicit parallelism that is inherent in the application. Second, WSC software exhibit significant workload churn. Users of Internet services are isolated from the service's implementation details by relatively well-defined and stable high-level APIs (e.g., simple URLs), making it much easier to deploy new software quickly. For example, key pieces of Google's services have release cycles on the order of days, compared to months or years for desktop software products. The datacenter is also a more homogeneous environment than the desktop. Large Internet services operations typically deploy a small number of hardware and system software configurations at any given point in time. Any heterogeneity arises primarily from the incentives to deploy more cost-efficient components that become available over time. Finally, although it may be reasonable for desktop-class software to assume a fault-free hardware operation for months or years, this is not true for datacenter-level services; Internet services must work in an environment where faults are part of daily life. Ideally, the cluster-level system software should provide a layer that hides most of that complexity from application-level software, although that goal may be difficult to accomplish for all types of applications.

The complexity of the raw WSC hardware as a programming platform can lower programming productivity because every new software product must efficiently handle data distribution, fault detection and recovery, and work around performance discontinuities (such as the DRAM/ disk gap and networking fabric topology issues mentioned earlier). Therefore, it is essential to produce software infrastructure modules that hide such complexity and can be reused across a large segment of workloads. Google's MapReduce, GFS, BigTable, and Chubby are examples of the kind of software that enables the efficient use of WSCs as a programming platform. With the introduction of accelerators in the fleet, similar software modules, such as Tensorflow, are needed to hide complexity there as well.

8.3   ECONOMICS AND ENERGY EFFICIENCY

The economics of Internet services demands very cost efficient computing systems, rendering it the primary metric in the design of WSC systems. Cost efficiency must be defined broadly to account for all the significant components of cost including facility capital and operational expenses (which include power provisioning and energy costs), hardware, software, management personnel, and repairs.

Power- and energy-related costs are particularly important for WSCs because of their size. In addition, fixed engineering costs can be amortized over large deployments, and a high degree of automation can lower the cost of managing these systems. As a result, the cost of the WSC "enclosure" itself (the datacenter facility, the power, and cooling infrastructure) can be a large component of its total cost, making it paramount to maximize energy efficiency and facility utilization. For example, intelligent power provisioning strategies such as peak power oversubscription may allow more computing to be deployed in a building.

The utilization characteristics of WSCs, which spend little time fully idle or at very high load levels, require systems and components to be energy efficient across a wide load spectrum, and particularly at low utilization levels. The energy efficiency of servers and WSCs is often overestimated using benchmarks that assume operation peak performance levels. Machines, power conversion systems, and the cooling infrastructure often are much less efficient at the lower activity levels, for example, at 30% of peak utilization, that are typical of production systems. We suggest that *energy proportionality* be added as a design goal for computing components. Ideally, energy-proportional systems will consume nearly no power when idle (particularly while in active idle states) and grad-ually consume more power as the activity level increases.

Energy-proportional components could substantially improve energy efficiency of WSCs without impacting the performance, availability, or complexity. Since the publication of the first version of this book, CPUs have improved their energy proportionality significantly while the re-maining WSC components have witnessed more modest improvements.

In addition, traditional datacenters themselves are not particularly efficient. A building's power utilization efficiency (PUE) is the ratio of total power consumed divided by useful (server) power; for example, a datacenter with a PUE of 2.0 uses an additional 1 W of power for every watt of server power. Unfortunately, many legacy datacenter facilities run at PUEs of 2 or greater, and PUEs of 1.5 are rare. Clearly, significant opportunities for efficiency improvements exist not just at the server level but also at the building level, as was demonstrated by Google's annualized 1.11 PUE across all its custom-built facilities as of late 2018 [GDCa].

Energy efficiency optimizations naturally produce lower electricity costs. However, power provisioning costs, that is, the cost of building a facility capable of providing and cooling a given level of power, can be even more significant than the electricity costs themselves—in Chapter 6 we showed that datacenter-related costs can constitute more than half of total IT costs in some de-ployment scenarios. Maximizing the usage of a facility's peak power capacity while simultaneously reducing the risk of exceeding it is a difficult problem but a very important part of managing the costs of any large-scale deployment.

8.4   BUILDING RESPONSIVE LARGE-SCALE SYSTEMS

8.4.1   CONTINUALLY EVOLVING WORKLOADS

In spite of their widespread adoption, in many respects, Internet services are still in their infancy as an application area. New products appear and gain popularity at a very fast pace with some of the services having very different architectural needs than their predecessors. For example, at the time of the first edition of this book, web search was the poster child for internet services. Video sharing on YouTube exploded in popularity in a period of a few months and the needs of such an application were distinct from earlier Web services such as email or search.

More recently, machine learning has exploded in its application across a wide variety of work-loads and use-cases. While machine learning is pervasive in multiple web services, some notable recent examples include: more meaning extraction in transitioning from web search to knowledge graphs, automatic image recognition and classification in photo and video sharing applications, and smart reply and automatic composition features in gmail. Beyond web services, machine learning is also transforming entire industries from health care to manufacturing to self-driving cars. Once again, this has led to a fundamental change in the computation needs for the WSCs that power these workloads. Now with adoption of Cloud Computing, as discussed in Chapter 2, we are in the early stages of yet another evolution in workloads. A particularly challenging consideration is that many parts of WSC designs include components (building, power, cooling) that are expected to last more than a decade to leverage the construction investment. The mismatch between the time scale for radical workload behavior changes and the design and life cycles for WSCs requires creative solutions from both hardware and software systems.

8.4.2   AMDAHL'S CRUEL LAW

Semiconductor trends suggest that future performance gains will continue to be delivered mostly by providing more cores or threads, and not so much by faster CPUs. That means that large-scale systems must continue to extract higher parallel efficiency (or speed-up) to handle larger, more interesting computational problems. This is a challenge today for desktop systems but perhaps not as much for WSCs, given the arguments we have made earlier about the abundance of thread-level parallelism in its universe of workloads. Having said that, even highly parallel systems abide by Amdahl's law, and there may be a point where Amdahl's effects become dominant even in this domain, limiting performance scalability through just parallelism. This point could come earlier; for example, if high-bandwidth, high-port count networking technology continues to be extremely costly with respect to other WSC components.

8.4.3   THE ATTACK OF THE KILLER MICROSECONDS

As introduced by Barroso et al [Bar+17], the "*killer microsecond problem*" arises due to a new breed of low-latency IO devices ranging from datacenter networking to accelerators, to emerging non-volatile memories. These IO devices have latencies on the order of microseconds rather than milliseconds. Existing system optimizations, however, are typically targeted at the nanosecond scale (at the computer architecture level) or millisecond scale (operating systems). Today's hardware and system software make an inadequate platform for microsecond-scale IO, particularly given the tension between the support for synchronous programming models for software productivity, and performance. New microsecond-optimized systems stacks, across hardware and software, are therefore needed. Such optimized designs at the microsecond scale, and corresponding faster IO, can in turn enable a virtuous cycle of new applications that leverage low latency communication, dramatically increasing the effective computing capabilities of WSCs.

8.4.4   TAIL AT SCALE

Similar to how systems need to be designed for fault tolerance, a new design constraint unique to WSCs is the design for *tail tolerance* [DB13]. This addresses the challenges in performance when even infrequent high-latency events (unimportant in moderate-size systems) can come to dominate overall service performance at the WSC level. As discussed in Chapter 2, for a system with a typical latency of 10 ms, but a 99th percentile of one second, the number of user requests that take more than one second goes from 1% to 63% when scaling from one machine to a cluster of 100 machines! Large online services need to be designed create a predictable and responsive (low latency) whole of out less predictable parts. Some broad principles that have been used in recent WSCs include prioritizing interactive requests, breaking tasks into finer-granularity units that can be interleaved to reduce head-of-line blocking, and managing background and one-off events carefully. Some specific software techniques used in Google systems are discussed in more detail in [DB13] including the use of canary requests and replicated hedge or speculative requests, and putting slow machines on "probation." Recent work on QoS management in WSCs discuss hardware support to improve tail tolerance as well [Mar+11, DK14, Lo+15]. The need for such class of techniques will only continue to be greater as the scale and complexity of WSCs increase.

8.5   LOOKING AHEAD

We are still learning how best to design and use this new class of machines, as they are still relatively nascent (~15 years) compared to traditional systems. Below, we identify some key challenges and opportunities in this space, based on our experience designing and using WSCs.

8.5.1   THE ENDING OF MOORE'S LAW

Overall, the broader computer architecture community faces an important and exciting challenge. As our thirst for computing performance increases, we must continue to find ways to ensure that performance improvements are accompanied by corresponding improvements in energy efficiency and cost efficiency. The former has been achieved in the past due to Dennard Scaling [Den+74]: every 30% reduction in transistor linear dimensions results in twice as many transistors per area and 40% faster circuits, but with a corresponding reduction to supply voltage at the same rate as transistor scaling. Unfortunately, this has become extremely difficult as dimensions approach atomic scales. It is now widely acknowledged that Dennard scaling has stopped in the past decade. This means that any significant improvements in energy efficiency in the foreseeable future are likely to come from architectural techniques instead of fundamental technology scaling.

More recently, we are also seeing trends that classic Moore's law—improvements in cost-efficiency—is also slowing down due to a number of factors spanning both economic considerations (e.g., fabrication costs in the order of billions of dollars) and fundamental physical limits (limits of CMOS scaling). This is a more fundamental disruption to the industry. The challenges are particularly exacerbated by our earlier discussion on evolving workloads and growing demand, with deeper analysis over ever growing volumes of data, new diverse workloads in the cloud, and smarter edge devices. Again, this means that continued improvements in computing performance need to come from architectural optimizations, for area and resource efficiency, but also around more hardware-software codesign and fundamental new architectural paradigms.

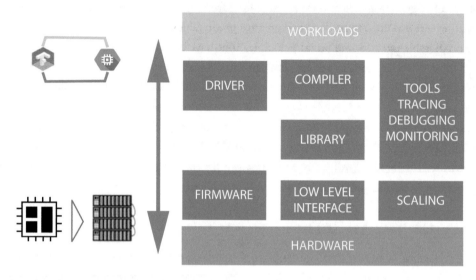

Figure 8.1: Accelerator ecosystems are important.

8.5.2 ACCELERATORS AND FULL SYSTEM DESIGNS

Accelerators are by far the most promising approach to addressing the end of Moore's Law. By tailoring the architecture to the application, we can achieve both improved power and area efficiency. The tradeoff often is that we sacrifice generality and flexibility to increase efficiency for specific types of workloads. In many respects, GPUs were the first and most successful example of the success of this approach, followed by other accelerators discussed in Chapter 3 such as FPGAs deployed by Microsoft and others, and full ASIC solutions such as Google's TPUs.

Accelerators also present a great example of how WSC design decisions need to evolve with changing workload requirements. In the first edition of this book, we argued *against* specialized computing pointing out that the promise of greater efficiency was not worth the tradeoffs of restricting the number of workloads that could benefit from them. However, since then, two key trends have changed that thinking. First, the slowing of Moore's law has made accelerators more appealing compared to general-purpose systems, but second, perhaps more importantly, deep learning models took off in a big way enabling specialized hardware in this space to power a broad spectrum of new machine learning solutions. As a result, earlier in this decade Google began to more broadly deploy GPUs but also initiated a program to build more specialized accelerators.

Accelerators in WSCs are still nascent, but our experience over the past few years have identified some key opportunities. First, hardware is just the proverbial tip of the iceberg (Figure 8.1).

Accelerator design needs to take a holistic view, across hardware and software for sure, but also across large-scale distributed system deployment. Our discussions earlier about the implications of designing at scale for general-purpose systems apply equally to the design of accelerators as well, illustrated by the design of pods of TPUs in Chapter 3. Additionally, similar to our discussions of the software stack in Chapter 2 for traditional systems, it is important to consider the full system stack for accelerators. This includes thinking about the design of accelerators in the context of the supporting interfaces and compilers, but also considering the tool and ecosystem support for tracing, debugging, monitoring, etc. A principled approach to the hardware-software codesign of accelerators is needed, one that carefully addresses problem decomposition, complexity encapsulation, modularity and interfaces, technical debt, and performance. A lot can be learned from our rich experience in software design [Ous18].

Beyond customization for specific workload classes like deep learning, search and video serving, there are also significant opportunities for the "long tail" of workloads. It is notable that nearly one out three compute cycles at Google is attributable to a handful of "datacenter tax" functions that cross-cut all applications. Similarly, the "killer microsecond" opportunities discussed earlier motivate new hardware innovations as well [Bar+17]. Storage, networking, and security are other broad areas where there are other opportunities for acceleration.

8.5.3 SOFTWARE-DEFINED INFRASTRUCTURE

Another promising approach for future WSCs is to embrace software-defined infrastructure. As discussed in prior chapters, WSCs have a lot of, often competing, constraints: how do we design at scale, at low costs, with ease of manageability and deployment, while also achieving high performance and reliability and efficiency? A software-defined infrastructure embraces a modular approach emphasizing efficiency in the design of the individual building blocks and focusing on capability through composability. This allows us to achieve the benefits of volume economics, fungibility, and agility with individual blocks, while allowing specialization, customization, and new capabilities at the broader composition layer.

SDN [Kre+14] is a good example where separating the network control plane (the policy engine) from the forwarding planes allows the the underlying infrastructure to be abstracted, while enabling the control to become more programmable and flexible. SDNs have been widely adopted in the the design of WSCs and in the broader industry [Jai+13, Kol14, Vah17]. We have an opportunity to consider similar software-defined approaches for the broader WSC design. Many ideas discussed in prior chapters including work around QoS management at the platform level [DK14, Lo+15, Mar+11], power management [Rag+08, Wu+16a], automatic memory and storage tier management [RC12], and more broadly disaggregated architectures [Lim+09, Gao+16] all play well into the theme of software-defined infrastructure.

It is also interesting to note that a software-defined infrastructure is the first step to enabling the greater adoption of ML-based automation in WSC designs. Early work in this space such as the use of ML-based models for prefetching [Has+18] or ML-based approaches to power management [EG16] are very promising, and we expect a lot more innovation in this space.

8.5.4   A NEW ERA OF COMPUTER ARCHITECTURE AND WSCS

John Hennessy and David Patterson, titled their 2018 Turing award lecture "A Golden Age for Computer Architecture." We think there is a similar golden age for WSCs coming as well.

Beyond accelerators and software-defined infrastructure, there are a few other exciting trends worth noting. Data is growing much faster than compute. We recently noted [Lot+18] that the pace of bytes of data uploaded to Youtube is exponentially diverging from the pace of traditional compute processing growth (an order of magnitude over the past decade). While there has been significant innovation in the WSC software stack around data storage and processing (for example, in prior chapters, we discussed GFS, Colossus, MapReduce, TensorFlow, etc.), such innovation has been relatively agnostic to the underlying hardware. Emerging new memory technologies such as Intel's 3D-Xpoint [3DX] or Samsung ZNAND [Sam17] present some fundamental technology disruptions in the memory/storage hierarchy. Co-designing new WSC hardware architectures for emerging data storage and processing needs will be an important area.

WSCs and cloud-based computation also offer the potential to reduce the environmental footprint of IT. On the server side, better utilization, lower PUE, and faster introduction of new hardware can significantly reduce the overall energy footprint as workloads move from inefficient on-premise deployments to cloud providers. On the client side, mobile devices don't need to store all data locally or process it on the device, leading to much more energy efficient clients as well.

Recent attacks like Spectre and Meltdown that exploit timing to inappropriately access data (discussed in Chapter 2) point to the growing importance of thinking of security as a first-class design constraint, motivating what some have called Architecture 2.0—a rethink of the hardware software interfaces to protect information [Hen+18]. At a system level, more hardware-software codesign for security is needed. Some examples include Google's Titan root of trust chip [Sav+17] or recent discussions around enclaves (e.g., Intel SGX).

One other opportunity for the broader community to counter the slowing of Moore's law is around *faster* hardware development. Moore's law is often formulated as improved performance over cost *over time* but the time variable does not get as much attention. If we can accelerate the cadence of introducing new hardware innovation in to the market, that can potentially offset slower performance increases per generation (sometimes referred to as optimizing the "area-under-the-curve"). How can we release "early and often" in WSC hardware akin to WSC software? Recent work in this area, for example around plug-and-play chiplets, post-silicon debugging, and ASIC clouds [ERI], as well as community efforts such as the RISC-V foundation (www.riscv.org) present interesting opportunities in this direction. Additional key challenges are around approaches to test and deploy custom hardware *at scale*. How do we "launch and iterate" at scale for hardware? Some preliminary work in this area include WSMeter [Lee+18] and FireSim [Kar+18] but more work is needed.

Looking further out, the growing proliferation of smart devices and increased computation at the edge motivate a rethink of end-to-end system tradeoffs. Much as WSCs changed how we think of datacenter design by drawing the box at defining the warehouse as a computer, thinking about interconnected WSCs and their relationship with computing at the edge and in the network will be important in the next decade.

8.6   CONCLUSIONS

Computation is moving into the cloud, and thus into WSCs. Software and hardware architects must be aware of the end-to-end systems to design good solutions. We are no longer designing individual "pizza boxes," or single-server applications, and we can no longer ignore the physical and economic mechanisms at play in a warehouse full of computers. At one level, WSCs are simple—just a few thousand servers connected via a LAN. In reality, building a cost-efficient massive-scale computing platform that has the necessary reliability and programmability requirements for the next generation of cloud-computing workloads is as difficult and stimulating a challenge as any other in computer systems today. We hope that this book will help computer scientists and practitioners understand and contribute to this exciting area.

Bibliography

[3DX]   3D XPoint™: A breakthrough in non-volatile memory technology. https://www.intel.com/content/www/us/en/architecture-and-technology/intel-micron-3d-xpoint-webcast.html. 74, 159, 166

[Abt+10]   D. Abts, M. R. Marty, P. M. Wells, P. Klausler, and H. Liu. 2010. Energy proportional datacenter networks. *SIGARCH Comput. Archit. News,* 38(3) (June 2010), 338–347. DOI: 10.1145/1816038.1816004. 115

[AF12]   D. Abts and B. Felderman. A guided tour of data-center networking. *Commun. ACM,* 55(6) (2012), p. 44–51. DOI: 10.1145/2184319.2184335. 65

[AI11]   ANSI INCITS: T11: Fibre Channel Interfaces. https://standards.incits.org/apps/group_public/workgroup.php?wg_abbrev=t11. 62

[AI462]   ANSI INCITS 462: Information technology, fibre channel: Backbone 5 (FC-BB-5). 62

[AK11]   D. Abts and J. Kim. *High-performance Datacenter Networks: Architectures, Algorithms, and Opportunities.* Morgan & Claypool, San Rafael, CA (2011). DOI: 10.2200/S00341ED1V01Y201103CAC014. 65

[AlF08]   M. Al-Fares, A. Loukissas, and A. Vahdat. A scalable, commodity datacenter network architecture. In *Proceedings of the ACM SIGCOMM 2008 Conference on Data Communication,* Seattle, WA, August 17–22 (2008).

[AM08]   D. Atwood and J. G. Miner. Reducing datacenter costs with an air economizer. IT@ Intel Brief (August 2008). 92

[AMD12]   AMD, SeaMicro product pages, http://www.seamicro.com/SM10000 (2012).

[Ana+11]   G. Ananthanarayanan, A. Ghodsi, S. Shenker, and I. Stoica. Disk-locality in datacenter computing considered irrelevant. In *Proceedings of the 13th Workshop on Hot Topics in Operating Systems (HotOS XIII)*, Napa, CA (May 2011). 70

[Ana+13]   R. Ananthanarayanan, V. Basker, S. Das, A. Gupta, H. Jiang, T. Qiu, A. Reznichenko, D. Ryabkov, M. Singh, and S. Venkataraman. Photon: Fault-tolerant and scalable joining of continuous data streams. *SIGMOD* (2013). https://static.googleusercontent.com/media/research.google.com/en//pubs/archive/41318.pdf. 69

[And+11] D. G. Andersen, J. Franklin, M. Kaminsky, A. Phanishayee, L. Tan, and V. Vasudevan. 2011. FAWN: a fast array of wimpy nodes. *Commun. ACM*, 54(7) (July 2011), 101–109. DOI: 10.1145/1965724.1965747. 53, 69

[Arm+10] M. Armbrust, A. Fox, R. Griffith, A. D. Joseph, R. Katz, A. Konwinski, G. Lee, D. Patterson, A. Rabkin, I. Stoica, and M. Zaharia. A view of cloud computing. *Commun. ACM*, 53(4) (April 2010), 50–58. DOI: 10.1145/1721654.1721672. 6

[Bah+07] P. Bahl, R. Chandra, A. Greenberg, S. Kandula, D. A. Maltz, and M. Zhang. Toward highly reliable enterprise network services via inference of multi-level dependencies. In *Proceedings of SIGCOMM* (2007). 31

[Bai+07] L. N. Bairavasundaram, G. R. Goodson, S. Pasupathy, and J. Schindler. An analysis of latent sector errors in disk drives. In *Proceedings of the 2007 ACM SIGMETRICS International Conference on Measurement and Modeling of Computer Systems*, San Diego, CA (June 12–16, 2007). SIGMETRICS '07. 149

[Bak+11] J. Baker, C. Bond, J. C. Corbett, J. J. Furman, A. Khorlin, J. S. Larson, J.-M. Leon, Y. Li, A. Lloyd, and V. Yushprakh. Megastore: Providing scalable, highly available storage for interactive services. In *Proceedings of the Conference on Innovative Data System Research (CIDR)* (2011), pp. 223–234. http://static.googleusercontent.com/external_content/ untrusted_dlcp/research.google.com/en/us/pubs/archive/36971.pdf. 69

[Bar+00] L. A. Barroso, K. Gharachorloo, R. McNamara, A. Nowatzyk, S. Qadeer, B. Sano, S. Smith, R. Stets, and B. Verghese. 2000. Piranha: a scalable architecture based on single-chip multiprocessing. *SIGARCH Comput. Archit. News*, 28(2) (May 2000), pp. 282–293. DOI: 10.1145/342001.339696. 53

[Bar+03a] P. Barham, B. Dragovic, K. Fraser, S. Hand, T. Harris, A. Ho, R. Neugebauer, I. Pratt, and A. Warfield. Xen and the art of virtualization. In *Proceedings of the Nineteenth ACM Symposium on Operating Systems Principles*, (SOSP '03). Bolton Landing, NY (October 19–22, 2003).

[Bar+03b] P. Barham, R. Isaacs, R. Mortier, and D. Narayanan. Magpie: online modeling and performance-aware systems. In *Proceedings of USENIX HotOS IX* (2003). 31

[Bar+17] L. A Barroso, M. Marty, D. Patterson, and P. Ranganathan. Attack of the killer microseconds. *Comm. of the ACM*, 60(4) (April 2017), pp. 48–54. DOI: 10.1145/3015146. 74, 162, 165

[Bar11] L. A. Barroso. Warehouse-scale computing: Entering the teenage decade. Plenary talk at the ACM Federated Computer Research Conference. In *Proceedings of the 38th An-*

nual International Symposium on Computer Architecture (ISCA '11). ACM, San Jose, CA (June 2011).

[BDH03] L. A Barroso, J. Dean, and U. Hölzle. Web search for a planet: the architecture of the Google cluster. *IEEE Micro* (April 2003). 45, 53

[BH07] L. A. Barroso and U. Hölzle. The case for energy-proportional computing. *IEEE Computer*, 40(12) (December 2007). 110

[BM06] C. Belady and C. Malone. Preliminary assessment from Uptime Institute: IDC data center of the future US server power spend for 2005 as a baseline($6bn); applied a cooling factor of 1; applied a .6 multiplier to US data for WW amount. Data Center Power Projection to 2014, 2006 ITHERM, San Diego, CA (June 2006). 103

[Bre+16] E. Brewer, L. Ying, L. Greenfield, R. Cypher, and T. T'so. *Disks for Data Centers*, Google (2016) pp. 1–16. https://ai.google/research/pubs/pub44830. DOI: 10.1109/4236.939450. 66

[Bre01] E. A. Brewer. Lessons from giant-scale services. *IEEE Internet Comput.*, 5(4) (July/ August 2001), pp. 46–55. 34, 142, 143

[Bur06] M. Burrows. The chubby lock service for loosely-coupled distributed systems. In *Proceedings of OSDI'06: Seventh Symposium on Operating System Design and Implementation*, Seattle, WA (November 2006). 17, 20

[Cha+01a] B. Chandra, M. Dahlin, L. Gao, and A. Nayate. End-to-end WAN service availability. In *Proceedings of the 3rd Conference on USENIX Symposium on Internet Technologies and Systems*, 3, San Francisco, CA (March 26–28, 2001).

[Cha+01b] B. Chandra, M. Dahlin, L. Gao, A.-A. Khoja, A. Nayate, A. Razzaq, and A. Sewani. Resource management for scalable disconnected access to Web services. *Proceedings of the 10th International Conference on World Wide Web*, Hong Kong (May 1–5, 2001), pp. 245–256. DOI: 10.1145/371920.372059. 143

[Cha+01c] J. Chase, D. Anderson, P. Thakar, A. Vahdat, and R. Doyle. Managing energy and server resources in hosting centers. In *Proceedings of the ACM Symposium on Operating Systems Principles* (SOSP), Banff, Alberta, Canada (June 2001). 123

[Cha+06] F. Chang, J. Dean, S. Ghemawat, W. C. Hsieh, D. A. Wallach, M. Burrows, T. Chandra, A. Fikes, and R. E. Gruber. Bigtable: a distributed storage system for structured data. In *Proceedings of OSDI* (2006), Seattle, WA (2004). 9, 17, 20

[Cha+10] C. Chambers, A. Raniwala, F. Perry, S. Adams, R. Henry, R. Bradshaw, and Nathan. FlumeJava: Easy, efficient data-parallel pipelines. *ACM SIGPLAN Conference on Pro-*

gramming Language Design and Implementation (PLDI), pp. 363-375 (2010). https://ai.google/research/pubs/pub35650. 20, 68

[Che+05] Y. Chen, A. Das, W. Qin, A. Sivasubramaniam, Q. Wang, and N. Gautam. Managing server energy and operational costs in hosting centers. In *Proceedings of the ACM SIGMETRICS '05*, Banff, Alberta, Canada (June 2005). 123

[Che+07] G. Chen, W. He, J. Liu, S. Nath, L. Rigas, L. Xiao, and F. Zhao. Energy aware server provisioning and load dispatching for connection-intensive Internet services. *Microsoft Research Technical Report* MSR-TR-2007-130 (2007). 123

[Che+16] J. Chen, R. Monga, S. Bengio, and R. Jozefowicz. Revisiting distributed synchronous SGD. *International Conference on Learning Representations Workshop Track* (2016). https://arxiv.org/abs/1604.00981. 28

[Chr+10] K. Christensen, P. Reviriego, B. Nordman, M. Bennett, M. Mostowfi, and J. A. Maestro. IEEE 802.3az: the road to energy efficient Ethernet. *Comm. Mag., IEEE*, 48(11), (November 2010) pp. 50-56. DOI: 10.1109/MCOM.2010.5621967. 115

[Cis17] Cisco Public White Paper. Cisco Visual Networking Index: Forecast and Methodology, 2016–2021 (June, 6, 2017).

[Cla15] Clark, J. Google turning its lucrative web search over to AI machines. *Bloomberg Technology*, (October 26, 2015) www.bloomberg.com. 28

[Cli] Climate savers computing efficiency specs. http://www.climatesaverscomputing.org/about/tech-specs. 106

[Col17] L. Columbus. Cloud computing market projected to reach $411B by 2020. *Forbes* (October, 18, 2017). 1

[Cor+12] J. C. Corbett et al. Spanner: Google's globally-distributed database. *Proceedings of OSDI'12: Tenth Symposium on Operating System Design and Implementation*, Hollywood, CA (October 2012). 9, 17, 20, 69

[Coy05] E. F. Coyle. Improved muscular efficiency displayed as tour de France champion matures. *J. Appl. Physiolo.* (March 2005). DOI: 10.1152/japplphysiol.00216.2005.

[CPB03] E. V. Carrera, E. Pinheiro, and R. Bianchini. Conserving disk energy in network servers. In *Proceedings of the 17th Annual International Conference on Supercomputing*, (ICS '03). San Francisco, CA (June 23–26, 2003). 114

[CTX2] Cavium ThunderX2® ARM Processors, High Performance ARMv8 Processors for Cloud and HPC Server Applications. https://www.cavium.com/product-thunderx2-arm-processors.html. 55

[Dal+18] M. Dalton, D. Schultz, J. Adriaens, A. Arefin, A. Gupta, B. Fahs, D. Rubinstein, E. Cauich Zermeno, E. Rubow, J. A. Docauer, J. Alpert, J. Ai, J. Olson, K. DeCabooter, M. de Kruijf, N. Hua, N. Lewis, N. Kasinadhuni, R. Crepaldi, S. Krishnan, S. Venkata, Y. Richter, U. Naik, and A. Vahdat. Andromeda: Performance, isolation, and velocity at scale in cloud network virtualization. *NSDI* (2018). 42, 66

[DB13]   J. Dean and L. A. Barroso. The tail at scale. *Commun. ACM*, 56(2) (February 2013), pp. 80–86. DOI: 10.1145/2408776.2408794. 38, 162, 163

[Dea]   J. Dean. Machine learning for systems and systems for machine learning. In *Proceedings of the NIPS*. http://learningsys.org/nips17/assets/slides/dean-nips17.pdf. 58

[Dea+12] J. Dean et al. Large scale distributed deep networks. *NIPS* (2012). https://ai.google/research/pubs/pub40565. 28

[Dea09]   J. Dean. Design lessons and advice from building large scale distributed systems, *LADIS* (2009). https://perspectives.mvdirona.com/2009/10/jeff-dean-design-lessons-and-advice-from-building-large-scale-distributed-systems/. 40

[DEC]   Dell Energy Calculator. http://www.dell.com/calc. 121

[DeC+07] G. DeCandia, D. Hastorun, M. Jampani, G. Kakulapati, A. Lakshman, A. Pilchin, S. Sivasubramanian, P. Vosshall, and W. Vogels. Dynamo: Amazon's highly available key-value store. In *Proceedings of the 21st ACM Symposium on Operating Systems Principles*, Stevenson, WA (October 2007). DOI: 10.1145/1323293.1294281. 9, 17, 20, 68

[Del97]   T. J. Dell. A white paper on the benefits of chipkill-correct ECC for PC server main memory," *IBM Microelectronics Division* (Rev. 11/19/1997). 148

[Den+74] R. Dennard et al. Design of ion-implanted MOSFETs with very small physical dimensions. *IEEE J. Solid State Circuits* SC-9, 5 (October 1974), pp. 256–268. DOI:10.1109/JSSC.1974.1050511. 163

[DG08]   J. Dean and S. Ghemawat. MapReduce: simplified data processing on large clusters. *Commun. ACM*, 51(1) (2008), pp. 107–113. DOI: 10.1145/1327452.1327492. 17, 20

[DK14]   C. Delimitrou and C. Kozyrakis. Quasar: resource-efficient and QoS-aware cluster management. In *Proceedings of the 19th International Conference on Architectural Support for Programming Languages and Operating Systems* (ASPLOS '14). ACM, New York, (2014) pp. 127–144. DOI: 10.1145/2541940.2541941. 117, 163, 165

[DMDC]   Dell Modular Data Center: http://www.dell.com/learn/us/en/555/shared-content~data-sheets~en/documents~hyper-efficient-modular-data-center.pdf.

[DuP07]   Dupont Fabros Technology Inc. SEC Filing (S-11) 333-145294, August 9, 2007.

[DuP11]   DuPont Fabros Technology Annual Report 2011. http://www.dft.com/sites/default/files/2011-10k.pdf. 130

[DWG92]   D. DeWitt and J. Gray. Parallel database systems: The future of high performance database processing," *CACM*, 36(6) (June 1992). 54

[Dye06]   D. Dyer. Current trends/challenges in datacenter thermal management—a facilities perspective. Presentation at *ITHERM*, San Diego, CA (June 1, 2006). 87

[eBay12]   eBay Containers Push Frontiers of Efficiency: http://www.datacenterknowledge.com/archives/2012/02/27/ebay-containers-push-frontiers-of-efficiency/. 96

[EG16]   R. Evans and J. Gao. DeepMind AI reduces Google data centre cooling bill by 40%. https://deepmind.com/blog/deepmind-ai-reduces-google-data-centre-cooling-bill-40/ (July 20, 2016). 103, 166

[EPA07]   U.S. Environmental Protection Agency. Report to Congress on server and datacenter energy efficiency. *Public Law* 109–431 (August 2, 2007). 76

[ERI]   Beyond scaling: An electronics resurgence initiative. *DARPA*, (2017). https://www.darpa.mil/news-events/2017-06-01. 166

[ES+]   N. El-Sayed, I. Stefanovici, G. Amvrosiadis, A. A. Hwang, and B. Schroeder. Temperature management in data centers: Why some (might) like it hot. *SIGMETRICS'12*. http://www.cs.toronto.edu/~bianca/papers/temperature_cam.pdf. 104, 149

[ES04]   J. Elerath and S. Shah. Server class disk drives: how reliable are they? *IEEE Reliability and Maintainability, 2004 Annual Symposium—RAMS* (January 2004).

[FF04]   M. E. Femal and V. W. Freeh. Safe overprovisioning: using power limits to increase aggregate throughput. In *4th International Workshop on Power Aware Computer Systems* (PACS 2004) (December 2004), pp. 150–164. 123

[FF05]   M. E. Femal and V. W. Freeh. Boosting datacenter performance through non-uniform power allocation. In *Proceedings of the Second International Conference on Automatic Computing*, (June 13–16, 2005), pp. 250–261. 20, 123

[Fir+18]   D. Firestone, A. Putnam, S. Mundkur, D. Chiou, A. Dabagh, M. Andrewartha, H. Angepat, V. Bhanu, A. M. Caulfield, E. S. Chung, H. Kumar Chandrappa, S. Chaturmohta, M. Humphrey, J. Lavier, N. Lam, F. Liu, K. Ovtcharov, J. Padhye, G. Popuri, S. Raindel, T. Sapre, M. Shaw, G. Silva, M. Sivakumar, N. Srivastava, A. Verma, Q. Zuhair, D. Bansal, D. Burger, K. Vaid, D. A. Maltz, and A. G. Greenberg. Azure Accelerated Networking: SmartNICs in the Public Cloud. NSDI 2018. 66

[Fit+03]   B. Fitzpatrick et al. (2003). Memcached. http://memcached.org/about. 69

[FM11]   S. H. Fuller and L. I. Millet (eds) *The Future of Computing Performance: Game Over or Next Level*. National Academies Press; Committee on Sustaining Growth in Computing Performance, National Research Council (2011). http://www.nap.edu/catalog.php?record_id=12980. 106

[Fon+07]   R. Fonseca, G. Porter, R. H. Katz, S. Shenker, and I. Stoica. X-trace: a pervasive network tracing framework. In *Proceedings of 4th USENIX Symposium on Networked Systems Design & Implementation*, (2007), pp. 271–284. 20

[For+10]   D. Ford, F. Labelle, F. I. Popovici, M. Stokely, V.-A. Truong, L. Barroso, C. Grimes, and S. Quinlan. Availability in globally distributed storage systems. In *Proceedings of the 9th USENIX Sympsium on Operating Systems Design and Implementation* (OSDI'10) (October 2010), Vancouver, BC, Canada. http://static.usenix.org/event/osdi10/tech/full_papers/Ford.pdf. 68, 144

[FWB07]   X. Fan, W. Weber, and L. A. Barroso. Power provisioning for a warehouse-sized computer. In *Proceedings of the 34th Annual International Symposium on Computer Architecture*, ISCA '07, San Diego, CA (June 9–13, 2007). DOI: /10.1145/1250662.1250665. 31, 110, 121, 122, 152

[Gan+]   A. Gandhi, M. Harchol-Balter, R. Das, J. O. Kephart, and C. Lefurgy. Power capping via forced idleness. http://repository.cmu.edu/compsci/868/. 115

[Gao+16]   P. Gao, A. Narayan, S. Karandikar, J. Carreira, S. Han, R. Agarwal, S. Ratnasamy, and S. Shenker. Network requirements for resource disaggregation. *OSDI* (2016). https://dl.acm.org/citation.cfm?id=3026897. 165

[GCAML]   Google Cloud AutoML. https://cloud.google.com/automl/. 59

[GCE]   Google Compute Engine SLAs: https://cloud.google.com/compute/sla. 155

[GCP]   Google Cloud Platform. https://cloud.google.com/. 42, 155

[GCRZ]   Google Cloud: Regions and Zones. https://cloud.google.com/compute/docs/regions-zones/. 155

[GDCa]   Google Data centers. http://www.google.com/about/datacenters/efficiency/internal/index.html#measuring-efficiency. 101, 161

[GDCb]   Google Data centers. Efficiency: How we do it. http://www.google.com/about/datacenters/efficiency/internal/index.html#measuring-efficiency. 103

[GDCC]   An Overview: Global Data Center Construction. https://wiredre.com/an-overview-global-data-center-construction/. 130

[GF]   A. J. George and G. Ferrand. Cost study of AC vs DC data center power topologies based on system efficiency. Eltek. https://tinyurl.com/yaemk37v. 82

[GGL03]   S. Ghemawat, H. Gobioff, and S-T. Leung. The Google file system, In *Proceedings of the 19th ACM Symposium on Operating Systems Principles*, Lake George, NY (October 2003). 20, 68, 109, 140

[GGr]   Google green, The Big Picture: http://www.google.com/green/bigpicture/#/. 101

[Gha+17] C. Ghali, A. Stubblefield, E. Knapp, J. Li, B, Schmidt, J. Boeuf. Application Layer Transport Security, (2017), https://cloud.google.com/security/encryption-in-transit/application-layer-transport-security/. 44

[GInc09] Google Inc.. Efficient Data Center Summit, April 2009. http://www.google.com/about/datacenters/efficiency/external/2009-summit.html. 96, 105

[GInca]   Google Inc. Efficient computing—step 2: efficient datacenters. http://www.google.com/corporate/green/datacenters/step2.html.

[GIncb]   Google Inc. Efficient Data Center, Part 1. http://www.youtube.com/watch?v=Ho1GEyftpmQ starting at 0:5930.

[GMT06]   S. Greenberg, E. Mills, and B. Tschudi. Best practices for data centers: lessons learned from benchmarking 22 datacenters. *2006 ACEEE Summer Study on Energy Efficiency in Buildings*. http://evanmills.lbl.gov/pubs/pdf/aceee-datacenters.pdf. 104

[Gov+]   S. Govindan, D. Wang, A. Sivasubramaniam, and B. Urgaonkar. Leveraging stored energy for handling power emergencies in aggressively provisioned datacenters. http://csl.cse.psu.edu/publications/asplos12.pdf. DOI: 10.1145/2189750.2150985. 125

[Gra90]   J. Gray. A census of tandem system availability between 1985 and 1990. *Tandem Technical Report* 90.1 (January 1990). 144

[GStaDb] Google Stackdriver Debugger. https://cloud.google.com/debugger/. 31

[GStaDr] Google Stackdriver. https://cloud.google.com/stackdriver/. 30, 31

[GStaPr] Google Stackdriver Profiler. https://cloud.google.com/profiler/. 31

[GStaTr] Google Stackdriver Trace. https://cloud.google.com/trace/.

[Hadoo]   The Hadoop Project. http://hadoop.apache.org. 17

[Ham07]   J. Hamilton. On designing and deploying internet-scale services. In *Proceedings of USENIX LISA* (2007). 34, 139

[Ham09] J. Hamilton. Cooperative expendable micro-slice servers (CEMS): low cost, low power servers for internet-scale services. In *Proceedings of the 4th Biennial Conference on Innovative Data Systems Research* (CIDR), Asilomar, CA (January 4–7, 2009). 53

[Has+18] M. Hashemi, K. Swersky, J. A. Smith, G. Ayers, H. Litz, J. Chang, C. Kozyrakis, and P. Ranganathan. Learning memory access patterns. *ICML* (2018) 1924-1933. 166

[He+15] K. He, X. Zhang, S. Ren, and J. Sun. *Deep Residual Learning for Image Recognition, 2015.* https://arxiv.org/abs/1512.03385. 28

[Hea+06] T. Heath, A. P. Centeno, P. George, L. Ramos, Y. Jaluria, and R. Bianchini. Mercury and freon: temperature emulation and management for server systems. In *Proceedings of the ACM International Conference on Architectural Support for Programming Languages and Operating Systems*, San Jose, CA (October 2006). DOI: 10.1145/1168918. 123

[Hen+18] J. Hennessy, P. Turner, J. Masters, and M. Hill. Hot18: Spectre/Meltdown and the era of security. Keynote at *Hotchips* (2018). 42, 44, 166

[Hes14] K. Heslin. Uptime Institute: 2014 Data Center Industry Survey. November 2014. https://journal.uptimeinstitute.com/2014-data-center-industry-survey/. 100

[Hin+11] Benjamin Hindman et al. Mesos: A platform for fine-grained resource sharing in the data center. In *NSDI'11: Proceedings of the 8th USENIX Symposium on Networked Systems Design and Implementation*, Boston, MA, USENIX Association (March, 2011). 117

[Höl10] U. Hölzle. Brawny cores still beat wimpy cores, most of the time. *IEEE MICRO* (July/August 2010). 54

[Höl12] U. Hölzle. OpenFlow @ Google. *Open Networking Summit*, Santa Clara (April 2012). http://www.opennetsummit.org/archives/apr12/hoelzle-tue-openflow.pdf. 65

[Höl16] U. Hölzle. Open compute project (OCP). *Summit 2016 Keynote*, San Jose CA.

[HPM13] HP Moonshot System: A new style of IT accelerating innovation at scale, Technical white paper (2013). https://goo.gl/3UGa4k. 54

[HPPC] HP Power Calculator. http://www.hp.com/configurator/powercalcs.asp. 121

[HPPOD] HP Performance Optimized Datacenter (POD). http://h18004.www1.hp.com/products/servers/solutions/datacentersolutions/pod/index.html.

[HSS12] A. A. Hwang, I. Stefanovici, and B. Schroeder. Cosmic rays don't strike twice: Understanding the nature of DRAM errors and the implications for system design, *ASPLOS 2012* (March 2012). 149

[Hsu+15] C. H. Hsu, Y. Zhang, M. A. Laurenzano, D. Meisner, T. Wenisch, J. Mars, L. Tang, and R. G. Dreslinski. Adrenaline: Pinpointing and reining in tail queries with quick voltage boosting. *IEEE 21st International Symposium on High Performance Computer Architecture* (HPCA), Burlingame, CA (2015), pp. 271–282. 123

[IBC15]   The 2015 International Building Code. https://codes.iccsafe.org/public/document/IBC2015/chapter-6-types-of-construction. 77

[Intel18] *Intel® 64 and IA-32 Architectures Software Developer's Manual Volume 3B: System Programming Guide, Part 2.* May 2018. https://software.intel.com/en-us/articles/intel-sdm. 116

[IntTu]   Intel® Turbo Boost Technology 2.0. https://www.intel.com/content/www/us/en/architecture-and-technology/turbo-boost/turbo-boost-technology.html. 118

[IntXe]   Intel® Xeon® Platinum 8164 Processor. https://ark.intel.com/products/120503/Intel-Xeon-Platinum-8164-Processor-35_75M-Cache-2_00-GHz. 118

[Isa+07] M. Isard, M. Budiu, Y. Yu, A. Birrell, and D. Fetterly. Dryad: distributed data-parallel programs from sequential building blocks. In *Proceedings of the 2nd ACM Sigops/Eurosys European Conference on Computer Systems 2007*, Lisbon, Portugal (March 21–23, 2007). 17

[Isa07]   M. Isard. Autopilot: automatic datacenter management. *SIGOPS Operating Systems Review*, 41(2) (April 2007), pp. 60–67. DOI: 10.1145/1243418.1243426. 20

[iSC03]   Internet Small Computer Systems Interface (iSCSI). The Internet Society: https://tools.ietf.org/html/rfc3720. 62

[Jai+13] S. Jain et al. B4: Experience with a globally-deployed software defined WAN. *SIG-COMM* (2013). 65, 165

[Jia+08] W. Jiang, C. Hu, Y. Zhou, and A. Kanevsky. Are disks the dominant contributor for storage failures?: A comprehensive study of storage subsystem failure characteristics. *Trans. Storage* 4, 3, Article 7 (November 2008), 25 pages. DOI: 10.1145/1416944.141694. 144

[Jou+17] N. P. Jouppi et al. In-datacenter performance analysis of a tensor processing unit. In *Proceedings of the 44th Annual International Symposium on Computer Architecture* (ISCA), Toronto (2017) pp. 1–12. DOI: 10.1145/3079856.3080246. 28, 56, 58, 119, 120

[Jou+18] N. Jouppi et al. Motivation for and evaluation of the first tensor processing unit. *IEEE Micro* (2018), pp 10–19. DOI: 10.1109/MM.2018.032271057. 27, 28

[Kan+15] S. Kanev, J. Pablo Darago, K. Hazelwood, P. Ranganathan, T. Moseley, G. Y. Wei, and D. Brooks. Profiling a warehouse-scale computer. *ACM SIGARCH Comput.Archit. News*, 43, pp. 158–169 (2015). DOI: 10.1145/2872887.2750392. 21

[Kar+18] S. Karandikar, H. Mao, D. Kim, D. Biancolin, A. Amid, D. Lee, N. Pemberton, E. Amaro, C. Schmidt, A. Chopra, Q. Huang, K. Kovacs, B. Nikolic, R. Katz, J. Bachrach, and K. Asanovic. FireSim: FPGA-accelerated cycle-exact scale-out system simulation in the public cloud. *2018 ACM/IEEE 45th Annual International Symposium on Computer Architecture* (ISCA), (2018), pp. 29–42. DOI: 10.1109/ISCA.2018.00014. 167

[Kas+15] H. Kasture, D. B. Bartolini, N. Beckmann, and D. Sanchez. Rubik: Fast analytical power management for latency-critical systems. *48th Annual IEEE/ACM International Symposium on Microarchitecture* (MICRO), Waikiki, HI (2015), pp. 598–610. 123

[KKI99] M. Kalyanakrishnam, Z. Kalbarczyk, and R. Iyer. Failure data analysis of a LAN of Windows NT based computers. *IEEE Symposium on Reliable Distributed Systems*, 0(0), p. 178, *18th IEEE Symposium on Reliable Distributed Systems* (1999). DOI: 10.1109/RELDIS.1999.805094. 145, 146

[Kol14] B. Koley. *Software Defined Networking at Scale*. 2014. https://ai.google/research/pubs/pub42948. 165

[Kon+12] V. Kontorinis, L. E. Zhang, B. Aksanli, J. Sampson, H. Homayoun, E. Pettis, D.M. Tullsen, and T. Simunic Rosing. Managing distributed ups energy for effective power capping in data centers. *SIGARCH Comput. Archit. News* 40(3) (June 2012), pp. 488–499. DOI: 10.1145/2366231.2337216. 125

[Koo+07] J. G. Koomey, K. Brill, P. Turner, J. Stanley, and B. Taylor. A simple model for determining true total cost of ownership for data centers. *Uptime Institute White Paper*, Version 2 (October, 2007). 129

[Koo+11] J.G. Koomey, S. Berard, M. Sanchez, and H. Wong. Implications of historical trends in the electrical efficiency of computing. *Ann. Hist Comput., IEEE*, 33(3), pp.46–54, (March 2011). DOI: 10.1109/MAHC.2010.28. 106

[Koo11] J. G. Koomey. *Growth in Data Center Electricity Use 2005 to 2010*. Analytics Press, 2011. 76

[Kre+14] D. Kreutz, F. M. V. Ramos, P. . Veríssimo, C. E. Rothenberg, S. Azodolmolky, and S. Uhlig. *Software-Defined Networking: A Comprehensive Survey*. OpenFlow Specification (2014). https://ieeexplore.ieee.org/abstract/document/6994333/. 165

[Kum+15] A. Kumar, S. Jain, U. Naik, N. Kasinadhuni, E. C. Zermeno, C. S. Gunn, J. Ai, B. Carlin, M. Amarandei-Stavila, M. Robin, A. Siganporia, S. Stuart, and A. Vahdat. BwE: Flexible, hierarchical bandwidth allocation for WAN distributed computing, *SIGCOMM*, (2015) https://ai.google/research/pubs/pub43838. 65

[Lee+18]   J. Lee, C. Kim, K. Lin, L. Cheng, R. Govindaraju, and J. Kim. WSMeter: A performance evaluation methodology for Google's production warehouse-scale computers. *USA* (March 2018), https://dl.acm.org/citation.cfm?id=3173196. 167

[Lim+08]   K. Lim, P. Ranganathan, J. Chang, C. Patel, T. Mudge, and S. Reinhardt. Understanding and designing new server architectures for emerging warehouse-computing environments. *International Symposium on Computer Architecture*, 0(0), pp. 315–326, (2008). DOI: 10.1109/ISCA.2008.37. 53, 54

[Lim+09]   K. Lim, J. Chang, T. Mudge, P. Ranganathan, S. K. Reinhardt, and T. F. Wenisch. Disaggregated memory for expansion and sharing in blade servers. *ISCA'09* (June 20–24, 2009), Austin, TX. https://web.eecs.umich.edu/~twenisch/papers/isca09-disaggregate.pdf. 165

[Lim+13]   K. Lim, D. Meisner, A. G. Saidi, P. Ranganathan, and T. Wenisch. Thin Servers with Smart Pipes: Designing SoC Accelerators for Memcached. In *Proceedings of the 40th ACM International Symposium on Computer Architecture* (ISCA), Tel Aviv, Israel (June 2013). 20, 55

[Lo+14]   D. Lo, L. Cheng, R. Govindaraju, L. A. Barroso, and C. Kozyrakis. Toward energy proportionality for large-scale latency-critical workloads. In *Proceeding of the 41st Annual International Symposium on Computer Architecture* (ISCA '14). IEEE Press, Piscataway, NJ, pp. 301–312. http://csl.stanford.edu/~christos/publications/2014.pegasus.isca.pdf. 116, 123

[Lo+15]   D. Lo, L. Cheng, R. Govindaraju, P. Ranganathan, and C. Kozyrakis. 2015. Heracles: improving resource efficiency at scale. In *Proceedings of the 42nd Annual International Symposium on Computer Architecture* (ISCA '15). ACM, New York, pp. 450–462. DOI: 10.1145/2749469.2749475. 117, 163, 165

[Lot+18]   A. Lottarini, A. Ramirez, J. Coburn, M. A. Kim, P. Ranganathan, D. Stodolsky, and M. Wachsler. vbench: Benchmarking video transcoding in the cloud. (2018). https://dl.acm.org/citation.cfm?id=3173207. 25, 166

[LPS10]   W. Lang, J. Patel, and S. Shankar. Wimpy Node Clusters: What about non-wimpy workloads? In *Proceedings of the Sixth International Workshop on Data Management on New Hardware* (June 2010). DOI: 10.1145/1869389.1869396.

[Man09]   M. Manos. The Capacity Problem. http://loosebolts.wordpress.com/2009/06/02/chiller-side-chats-the-capacity-problem/. 121

[Mar+11]   J. Mars, L. Tang, R. Hundt, K. Skadron, and M. L. Soffa. 2011. Bubble-Up: increasing utilization in modern warehouse scale computers via sensible co-locations. In *Proceedings of the 44th Annual IEEE/ACM International Symposium on Microarchitecture* (MICRO-44). ACM, New York, pp. 248–259. DOI: 10.1145/2155620.2155650. 117, 163, 165

[MB06]   C. Malone and C. Belady. Metrics to characterize datacenter & IT equipment energy use. In *Proceedings of the Digital Power Forum*, Richardson, TX (September 2006). 100

[McKQ09] M. K. McKusick and S. Quinlan. GFS: Evolution on fast-forward. *ACM Queue* (August 1, 2009). DOI: 10.1145/1594204.1594206. 68

[Mei+11]   D. Meisner, C. M. Sadler, L. A. Barroso, W.-D. Weber, and T. F. Wenisch. 2011. Power management of online data-intensive services. *SIGARCH Comput. Archit. News* 39(3) (June 2011), pp.319–330. DOI: 10.1145/2024723.2000103. 116

[Mel+10]   S. Melnik et al. Dremel: Interactive analysis of web-scale datasets. In *Proceeding of the 36th International Conference on Very Large Data Bases* (2010), pp. 330–339, 2010. 17

[MGW09] D. Meisner, B. T. Gold, and T. F. Wenisch. PowerNap: eliminating server idle power. *SIGPLAN* Not. 44(3) (March 2009) pp. 205–216. DOI: 10.1145/1508284.1508269. 115, 116

[Mic12]   Microsoft Expands Dublin Cloud Computing Hub. Data Center Knowledge. February 23rd, 2012. http://www.datacenterknowledge.com/archives/2012/02/23/microsoft-expands-dublin-cloud-computing-hub/. 130

[Micro]   Microsoft's Top 10 Business Practices for Environmentally Sustainable Data Center. http://www.microsoft.com/environment/our_commitment/articles/datacenter_bp.aspx. 96

[Mil12]   R. Miller. Facebook has spent $210 million on Oregon data center. *Data Center Knowledge*. (January 30, 2012). http://www.datacenterknowledge.com/archives/2012/01/30/facebookhas-spent-210-million-on-oregon-data-center/. 130

[MKK04] W. D. McArdle, F. I. Katch, and V. L. Katch. *Sports and Exercise Nutrition*, 2nd ed., LWW Publishers, 2004.

[MLP18]   MLPerf. https://mlperf.org/assets/static/media/MLPerf-User-Guide.pdf. (2018). 28

[Moo+05] J. Moore, J. Chase, P. Ranganathan, and R. Sharma. Making scheduling "cool": temperature-aware workload placement in datacenters. In *Proceedings of the Annual Conference on USENIX Annual Technical Conference*, Anaheim, CA (April 10–15, 2005). 123

182   BIBLIOGRAPHY

[MPO]   Microsoft's Project Olympus. https://www.opencompute.org/wiki/Server/ProjectOlympus. 54

[Mur+16]   N. R. Murphy, B. Beyer, C. Jones, and J. Petoff. *Site Reliability Engineering: How Google Runs Production Systems*, 1st ed. (April 2016). 32

[Mys+09]   R. N. Mysore, A. Pamboris, N. Farrington, N. Huang, P. Miri, S. Radhakrishnan, V. Subramanya, and A. Vahdat. PortLand: A scalable fault-tolerant layer 2 data center network fabric. Department of Computer Science and Engineering, University of California San Diego. *SIGCOMM'09* (August 17–21, 2009), Barcelona, Spain. 61

[Nel+]   D. Nelson, M. Ryan, S. DeVito, K. V. Ramesh, P. Vlasaty, B. Rucker, B. Da y Nelson, et al. The role of modularity in datacenter design. *Sun BluePrints Online*. http://www.sun.com/storagetek/docs/EED.pdf. 104

[NVM]   NVM Express. http://nvmexpress.org/resources/specifications/. 62

[OAI18]   OpenAI. *AI and Compute*. (May 2018). https://blog.openai.com/ai-and-compute/. 56, 57

[OCP11]   *Open Compute Project, Battery Cabinet Hardware* v1.0. http://opencompute.org/wp/wp-content/uploads/2011/07/DataCenter-Battery-Cabinet-Specifications.pdf. 80

[OGP03]   D. Oppenheimer, A. Ganapathi, and D. A. Patterson. Why do Internet services fail, and what can be done about it? In *Proceedings of the 4th Conference on USENIX Symposium on Internet Technologies and Systems*, Volume 4, Seattle, WA (March 26–28, 2003). 143

[ONF12]   Open Networking Foundation. Software-Defined Networking: The New Norm for Networks (April 2012). https://www.opennetworking.org/images/stories/downloads/sdn-resources/white-papers/wp-sdn-newnorm.pdf. 65

[Ous+09]   J. Ousterhout, P. Agrawal, D. Erickson, C. Kozyrakis, J. Leverich, D. Mazières, S. Mitra, A. Narayanan, G. Parulkar, M. Rosenblum, S. M. Rumble, E. Stratmann, and R. Stutsman. The case for RAMClouds: Scalable high-performance storage entirely in DRAM. *SIGOPS Operating Systems Review*, 43,(4) (December 2009), pp. 92–105. 69

[Ous+11]   J. Ousterhout, P. Agrawal, D. Erickson, C. Kozyrakis, J. Leverich, D. Mazières, S. Mitra, A. Narayanan, D. Ongaro, G. Parulkar, M. Rosenblum, S. M. Rumble, E. Stratmann, and R. Stutsman. 2011. The case for RAMCloud. *Commun. ACM* 54(7) (July 2011), 121–130. DOI: 10.1145/1965724.1965751.

[Ous18]   J. Ousterhout. *A Philosophy of Software Design*. Yaknyam Press (April 2018), 178 pages. 165

[Pat+]   C. Patel et al. Thermal considerations in cooling large scale high compute density datacenters. http://www.flomerics.com/flotherm/technical_papers/t299.pdf.

[Pat+02] D. A. Patterson, A. Brown, P. Broadwell, G. Candea, M. Chen, J. Cutler, P. Enriquez, A. Fox, E. Kiciman, M. Merzbacher, D. Oppenheimer, N. Sastry, W. Tetzlaff, J. Traupman, and N. Treuhaft. Recovery-oriented computing (ROC): motivation, definition, techniques, and case studies. *UC Berkeley Computer Science Technical Report UCB//CSD-02-1175* (March 15, 2002). 147

[Pat+05] C. Patel et al. Cost model for planning, development and operation of a datacenter. http://www.hpl.hp.com/techreports/2005/HPL-2005-107R1.pdf. 129

[PB]   L. Page and S. Brin. The anatomy of a large-scale hypertextual search engine. *http://infolab.stanford.edu/~backrub/google.html*. 23

[Ped12]   M. Pedram. Energy-efficient datacenters. Computer-aided design of integrated circuits and systems. *IEEE Trans. Comput-Aided Des Integr Circ Sys*. (October 2012). DOI: 10.1109/TCAD.2012.2212898. 123

[PF]   M. K. Patterson and D. Fenwick. The state of datacenter cooling. *Intel Corporation White Paper*. http://download.intel.com/technology/eep/data-center-efficiency/state-of-date-center-cooling.pdf. 89, 94

[PGE]   PG&E. High performance datacenters. http://hightech.lbl.gov/documents/DATA_CENTERS/06_DataCenters-PGE.pdf. 104

[Pik+05] R. Pike, S. Dorward, R. Griesemer, and S. Quinlan. Interpreting the data: parallel analysis with Sawzall. *Sci.Program. J.*, 13(4) (2005), pp. 227–298. 17

[Pra+06] A. Pratt, P. Kumar, K. Bross, and T. Aldridge. Powering compute platforms in high efficiency data centers. *IBM Technology Symposium*, 2006. 82

[PUE10]   Distribution of PUE Ratios. (January, 2010). http://www.energystar.gov/ia/partners/prod_development/downloads/DataCenters_GreenGrid02042010.pdf. 100

[Put+14] A. Putnam, A. Caulfield, E. Chung, D. Chiou, K. Constantinides, J. Demme, H. Esmaeilzadeh, J. Fowers, J. Gray, . Haselman, S. Hauck, S. Heil, A. Hormati, J. Y. Kim, S. Lanka, E. Peterson, A. Smith, J. Thong, P. Yi Xiao, D. Burger, J. Larus, G. P. Gopal, and S. Pope. *A Reconfigurable Fabric for Accelerating Large-Scale Datacenter Services*. IEEE Press (June 1, 2014). 56

[PWB07]   E. Pinheiro, W.-D. Weber, and L. A. Barroso. Failure trends in a large disk drive population. In *Proceedings of 5th USENIX Conference on File and Storage Technologies* (FAST 2007), San Jose, CA (February 2007). 14, 20, 104, 149, 150

[QC240] Qualcomm Centriq 2400 Processor. https://www.qualcomm.com/products/qualcomm-centriq-2400-processor. 55

[Rag+08] R. Raghavendra, P. Ranganathan, V. Talwar, Z. Wang, and X. Zhu. No "power" struggles: coordinated multi-level power management for the datacenter. In *Proceedings of the ACM International Conference on Architectural Support for Programming Languages and Operating Systems*, Seattle, WA (March 2008). DOI: 10.1145/1353536.1346289. 117, 123, 165

[Ran+06] P. Ranganathan, D. Irwin, P. Leech, and J. Chase. Ensemble-level power management for dense blade servers. *Conference: Computer Architecture*, February, 2006. ISCA '06. https://www.researchgate.net/publication/4244659_Ensemble-Level_Power_Management_for_Dense_Blade_Servers. 83

[RC12] P. Ranganathan, and J. Chang. (Re) Designing data-centric data centers. (2012). http://www.hpl.hp.com/discover2012/pdf/2012_IEEEMicro_DCDC.pdf. 165

[Ren+10] G. Ren, E. Tune, T. Moseley, Y. Shi, S. Rus, and R. Hundt. Google-wide profiling: A continuous profiling infrastructure for data centers. *IEEE Micro* 30(4) (July 2010), pp. 65– 79. DOI: 10.1109/MM.2010.68. 31

[Reu08] Reuters. Dupont Fabros Technology, Inc. reports third quarter results. (November 5, 2008).

[Rey+06a] P. Reynolds, C. Killian, J. L. Wiener, J. C. Mogul, M. A. Shah, and A. Vahdat. Pip: detecting the unexpected in distributed systems. In *Proceedings of USENIX NSDI* (2006). 31

[Rey+06b] P. Reynolds, J. L. Wiener, J. C. Mogul, M. K. Aguilera, and A. Vahdat. WAP5: black box performance debugging for wide-area systems. In *Proceedings of the 15th International World Wide Web Conference* (2006). 31

[RFG02] Robert Frances Group. Total cost of ownership for Linux in the enterprise (July 2002). http://www-03.ibm.com/linux/RFG-LinuxTCO-vFINAL-Jul2002.pdf. 132

[Riv+07] S. Rivoire, M. A. Shah, P. Ranganathan, C. Kozyrakis, and J. Meza. Models and metrics to enable energy-efficiency optimizations. *Computer* 40(12) (December 2007), pp. 39–48. 106

[Sae+17] A. Saeed, N. Dukkipati, V. Valancius, V. The Lam, C. Contavalli, and A. Vahdat. Carousel: Scalable traffic shaping at end hosts. *SIGCOMM* (2017). 66

[Sam17] Samsung Z-NAND SSD technology brief. Ultra-low latency with Samsung Z-NAND SSD. 2017. https://www.samsung.com/us/labs/pdfs/collateral/Samsung_Z-NAND_Technology_Brief_v5.pdf. 159, 166

[Sav+17] U. Savagaonkar, N. Porter, N. Taha, B. Serebrin, and N. Mueller. Google Cloud Platform Blog, Titan in depth: Security in plaintext (August 2017). https://cloudplatform.googleblog.com/2017/08/Titan-in-depth-security-in-plaintext.html. 166

[SAV07] SAVVIS press release. SAVVIS sells assets related to two datacenters for $200 million. (June 29, 2007). http://news.centurylink.com/SAVVIS-Sells-Assets-Related-to-Two-Data-Centers-for-200-Million.

[Sch+13] M. Schwarzkopf et al. Omega: flexible, scalable schedulers for large compute clusters. *EuroSys* (2013).

[Ser17] D. Serenyi. Keynote. *2nd Joint International Workshop on Parallel Data Storage & Data Intensive Scalable Intensive Computing Systems* (November 2017). 9, 20, 152 , 154

[SG07a] B. Schroeder and G. A. Gibson. Understanding failures in petascale computers. *J. Phys.: Conf. Ser.* 78 (2007). DOI: 10.1088/1742-6596/78/1/012022. 145

[SG07b] B. Schroeder and G. A. Gibson. Disk failures in the real world: what does an MTTF of 1,000,000 hours mean to you?. In *Proceedings of the 5th USENIX Conference on File and Storage Technologies* (February 2007). 14, 149

[SGS08] S. Sankar, S. Gurumurthi, and M. R. Stan. Intra-disk parallelism: an idea whose time has come. In *Proceedings of the ACM International Symposium on Computer Architecture*, (June 2008), pp. 303–314. DOI: 10.1145/1394609.1382147. 114

[She+16] A. Shehabi, S. Smith, D. Sartor, R. Brown, M. Herrlin, J. Koomey, E. Masanet, N. Horner, I. Azevedo, and W. Littner. U.S. Department of Energy: United States Data Center Usage Report (June 2016). 100

[Shu+13] J. Shute, R. Vingralek, B. Samwel, B. Handy, C. Whipkey, E. Rollins, M. Oancea, K. Littlefield, D. Menestrina, S. Ellner, J. Cieslewicz, I. Rae, T. Stancescu, and H. Apte. F1: A distributed SQL database that scales. *PVLDB* 6(11): 1068–1079 (2013). 69

[Sig+10] B. H. Sigelman et al. Dapper, a large-scale distributed systems tracing infrastructure. *Google Research Report*, http://research.google.com/pubs/pub36356.html. 2010. 31

[Sil+16] D. Silver, A. Huang, C.J. Maddison, A. Guez, L. Sifre, G. Van Den Driessche, J. Schrittwieser, I. Antonoglou, V. Panneershelvam, M. Lanctot, and S. Dieleman. Mastering the game of Go with deep neural networks and tree search. *Nature* 529 (7587) (2016). 28

[Sin+15] A. Singh, J. Ong, A. Agarwal, G. Anderson, A. Armistead, R. Bannon, S. Boving, G. Desai, B. Felderman, P. Germano, A. Kanagala, J. Provost, J. Simmons, E. Tanda, J. Wanderer, U. Hölzle, S. Stuart, and A. Vahdat. Jupiter rising: A decade of Clos topologies and centralized control in Google's datacenter network, *SIGCOMM* (2015). 62, 65, 152

[Sna] Snappy: A fast compressor/decompressor. https://google.github.io/snappy/. 40

[SNI11] Storage Networking Industry Association (SNIA) Emerald Power Efficiency Measurement Specification V1.0 (August 2011). 106

[SNIA] SNIA Green Storage Initiative. http://www.snia.org/forums/green/.

[Sou12] W. Sounders. Server efficiency: Aligning energy use with workloads. *Data Center Knowledge* (June 12, 2012). 114

[SPC12] *Storage Performance Council SPC Benchmark 2/Energy Extension* V1.4 (November 2012). 106

[SPEC] SPEC Power. http://www.spec.org/power_ssj2008/. 106

[SPV07] S. Siddha, V. Pallipadi, and A. Van De Ven. Getting maximum mileage out of tickless In *Proceedings of the Linux Symposium*, Ottawa, Ontario, Canada (June 2007). 109

[SPW09] B. Schroeder, E. Pinheiro, and W.-D. Weber. DRAM errors in the wild: a large-scale field study. To appear in *Proceedings of SIGMETRICS,* 2009. 104, 149

[Techa] Techarp.com, Intel desktop CPU guide. http://www.techarp.com/showarticle.aspx?art-no=337&pgno=6. 134

[Terra] Terrazon Semiconductor. Soft errors in electronic memory—a white paper. http://www.tezzaron.com/about/papers/soft_errors_1_1_secure.pdf. 148

[TF05] M. Ton and B. Fortenbury. High performance buildings: datacenters—server power supplies. Lawrence Berkeley National Laboratories and EPRI (December 2005).

[TG500] The Green 500. http://www.green500.org. 106

[TGGa] The Green Grid. A Framework for Data Center Energy Efficiency: https://tinyurl.com/yayave27.

[TGGb] The Green Grid. The Latest on The Green Grid's Productivity Research. https://tinyurl.com/y92umkyn. 99

[TGGc] The Green Grid datacenter power efficiency metrics: PUE and DCiE. https://tinyurl.com/d5jmst. 99

[TGGd] The Green Grid, WP#22-Usage and Public Reporting Guidelines for The Green Grid's Infrastructure Metrics PUE/DCiE. https://tinyurl.com/y9gwt4tk. 101, 102

[TGGe] The Green Grid, Quantitative analysis of power distribution configurations for datacenters. http://www.thegreengrid.org/gg_content/.

[TGGf]   The Green Grid. Seven strategies to improve datacenter cooling efficiency. http://www. thegreengrid.org/gg_content/.

[TIA]   Telecommunications Industry Association standards. http://global.ihs.com/tia_tele-com_infrastructure.cfm?RID=Z56&MID=5280. 75, 76

[TIA942]   TIA-942 Telecommunications Infrastructure Standard for Data Centers. http://www. adc.com/us/en/Library/Literature/102264AE.pdf Page 5.

[TPC]   Transaction Processing Performance Council. http://www.tpc.org. 45

[TPC07a]   TPC-C Executive Summary for the Superdome-Itanium2 (February 2007). 45

[TPC07b]   TPC-C Executive Summary for the ProLiant ML350G5 (September 2007). 45

[TS06]   W. Pitt Turner IV and J. H. Seader. Dollars per kW plus dollars per square foot are a better datacenter cost model than dollars per square foot alone. *Uptime Institute White Paper* (2006). 130

[TSB]   W. Pitt Turner IV, J. H. Seader, and K. G. Brill. Tier classifications define site infra-structure performance. *Uptime Institute White Paper.* 75

[Tsc+03]   W. F. Tschudi, T. T. Xu, D. A. Sartor, and J. Stein. High performance datacenters: a research roadmap. Lawrence Berkeley National Laboratory, Berkeley, CA (2003).

[UpI12]   Uptime Institute: Important to recognize the dramatic improvement in data center efficiency. http://blog.uptimeinstitute.com/2012/09/important-to-recognize-the-dra-matic-improvement-in-data-center-efficiency/. 100

[UpIOS]   Uptime Institute: Data Center Site Infrastructure Tier Standard: Operational Sustain-ability. http://uptimeinstitute.com/publications. 75, 76

[UpIP]   Uptime Institute Publications. http://uptimeinstitute.com/publications.

[UpIT]   Uptime Institute: Data Center Site Infrastructure Tier Standard: Topology. http:// uptimeinstitute.com/publications. 76

[Vah+10]   A. Vahdat, M. Al-Fares, N. Farrington, R. N. Mysore, G. Porter, and S. Radhakrish-nan. Scale-out networking in the data center. *IEEE Micro* 30(4):29-41 (2010). DOI: 10.1109/ MM.2010.72. 61, 65

[Vah17]   A. Vahdat. *ONS Keynote: Networking Challenges for the Next Decade.* The Linux Foundation (April 2017). http://events17.linuxfoundation.org/sites/events/files/slides/ ONS%20Keynote%20Vahdat%202017.pdf. 165

[Ver+15]   A. Verma, L. Pedrosa, M. Korupolu, D. Oppenheimer, E. Tune, and J. Wilkes. Large-scale cluster management at Google with Borg. In *Proceedings of the Tenth European*

Conference on Computer Systems (EuroSys '15). ACM, New York, (2015). Article 18, 17 pages. DOI: 10.1145/2741948.2741964. 117, 152

[VMware] VMware infrastructure architecture overview white paper. http://www.vmware.com/pdf/vi_architecture_wp.pdf. 18

[Vog08] W. Vogels. Eventually consistent. *ACM Queue* (October 2008). http://queue.acm.org/detail.cfm?id=1466448. 34, 37

[Wan+] D. Wang, C. Ren, A. Sivasubramaniam, B. Urgaonkar, and H. Fathy. *Energy Storage in Datacenters: What, Where, and How much?* DOI: 10.1145/2318857.2254780. 125

[Wbt] Wet bulb temperature. http://en.wikipedia.org/wiki/Wet-bulb_temperature. 92

[Whi+] W. Whitted, M. Sykora, K. Krieger, B. Jai, W. Hamburgen, J. Clidaras. D. L. Beaty, and G. Aigner. Data center uninterruptible power distribution architecture. http:// www.google.com/patents/US7560831. 80

[Wu+16a] Q. Wu, Q. Deng, L. Ganesh, C. H. Hsu, Y. Jin, S. Kumar, B. Li, J. Meza, and Y. J. Song. Dynamo: Facebook's Data Center-Wide Power Management System. *2016 ACM/ IEEE 43rd Annual International Symposium on Computer Architecture* (ISCA), Seoul, (2016), pp. 469–480. 118, 165

[Wu+16b] Y. Wu, M. Schuster, Z. Chen., Q. V. Le, M. Norouzi., W. Macherey, M. Krikun, Y. Cao, Q. Gao, K. Macherey, J. Klingner, A. Shah, M. Johnson, X. Liu, L. Kaiser, S. Gouws, Y. Kato, T. Kudo, H. Kazawa, K. Stevens, G. Kurian, N. Patil, W. Wang, C. Young, J. Smith, J. Riesa, A. Rudnick, O. Vinyals, G. Corrado, M. Hughes, and J. Dean. Google's neural machine translation system: Bridging the gap between human and machine translation (September 26, 2016) http://arxiv.org/abs/1609.08144. 28

Author Biographies

Luiz André Barroso has worked across several engineering areas including web search, software infrastructure, storage availability, energy efficiency, and hardware design. He was the first manager of Google's Platforms Engineering team, the group responsible for designing the company's computing platform, and currently leads engineering infrastructure for Google Maps. Prior to Google, he was a member of the research staff at Digital Equipment Corporation (later acquired by Compaq), where his group did some of the pioneering work on processor and memory system design for multi-core CPUs. He holds a Ph.D. in computer engineering from the University of Southern California and B.S/M.S. degrees in electrical engineering from the PUC, Rio de Janeiro. Luiz is a Google Fellow, a Fellow of the ACM, and a Fellow of the American Association for the Advancement of Science.

Urs Hölzle served as Google's first vice president of engineering and has been leading the development of Google's technical infrastructure since 1999. His current responsibilities include the design and operation of the servers, networks, datacenters, and software infrastructure that power Google's internal and external cloud platforms. He is also renowned for both his red socks and his free-range Leonberger, Yoshka (Google's original top dog). Urs grew up in Switzerland and received a master's degree in computer science from ETH Zürich and, as a Fulbright scholar, a Ph.D. from Stanford. While at Stanford (and then a start-up later acquired by Sun Microsystems), he invented fundamental techniques used in most of today's leading Java compilers. Before joining Google, he was a professor of computer science at the University of California, Santa Barbara. He is a Fellow of the ACM and AAAS, a member of the Swiss Academy of Technical Sciences and the National Academy of Engineering, and serves on the board of the US World Wildlife Fund.

Parthasarathy Ranganathan is the area tech lead for Google's computing and datacenter hardware. Prior to this, he was an HP Fellow and Chief Technologist at Hewlett Packard Labs, where he led their research on systems and datacenters. Partha has worked on several interdisciplinary systems projects with broad impact on both academia and industry, including widely used innovations in energy-aware user interfaces, heterogeneous multi-cores, power-efficient servers, accelerators, and disaggregated and data-centric data centers. He has published extensively and is a co-inventor on more than 100 patents. He has been named a top-15 enterprise technology rock star by Business Insider, one of the top 35 young innovators in the world by MIT Tech Review, and is a recipient of the ACM SIGARCH Maurice Wilkes award and Rice University's Outstanding Young Engineering Alumni award. Partha is currently a Google distinguished engineer and is also a Fellow of the IEEE and ACM.